THE TALE OF THE LOBSTER

A Practical Guide to the American Lobster, The Inshore and Offshore Lobster Fisheries, and the Hardy Souls That Pursue One of the Ocean's Finest Delicacies

By

Robert Delano Martin

© 2002 by Robert Delano Martin. All rights reserved.

No part of this book may be reproduced, stored in a retrieval system, or transmitted by any means, electronic, mechanical, photocopying, recording, or otherwise, without written permission from the author.

ISBN: 1-4033-4512-0 (e-book)
ISBN: 1-4033-4513-9 (Paperback)
ISBN: 1-4033-4514-7 (Hardcover)

Library of Congress Control Number: 2002108212

This book is printed on acid free paper.

Printed in the United States of America
Bloomington, IN

1st Books - rev. 10/29/02

Table of Contents

PREFACE vii
Chapter 1. The American Lobster –
  A Threatened Marine Resource? 1
Chapter 2. The Anatomy And Biology Of The American Lobster ... 15
Chapter 3. The Lobster's Processes of Reproduction 34
Chapter 4. Life On The Bottom 58
Chapter 5. The Lobster Fisherman 73
Chapter 6. Practices And Conditions That Influence
  The Lobster Fishery 130
Chapter 7. "The Feast" – Buying, Cooking,
  And Eating The Lobster 213
Chapter 8. Some Commonly Asked Questions About Lobsters 240
Chapter 9. A Dictionary of Important
  Lobster and Lobstering Terms 255
Chapter 10. Let's Go Fishing - A Lobstering Pictorial 279
Epilogue: Be Merciful Unto The Lobster 317

IN APPRECIATION

It is a supposition on my part that every writer has experienced moments of despair and discouragement in the process of getting the words down on paper in order to tell a story that he or she feels a need to be told. Any such despair and discouragement, in my case, has been overcome as a result of the encouragement and valued assistance received from my loving family. For this, therefore, I am deeply grateful.

In appreciation, this book is dedicated to my wife Virginia, my sons Robert and Donald Martin, my daughters Dorothy Martin and Deborah Melanson, my son-in-law Douglas Melanson, and my daughter-in-law Michele Martin. This book is further dedicated to my grandchildren Philip Melanson, and Jessica and Kyle Martin, who someday will be old enough to read it and hopefully gain some measure of knowledge from it.

A special word of thanks is extended to Douglas Melanson for his time, effort, and computer know-how in converting the written manuscript into the compact disc format preferred by the publisher.

To all of the above, I thank you very much.

Robert Delano Martin

The beautiful inner harbor at Rockport, Massachusetts on the tip of Cape Ann. In the foreground are lobster boats tied up at their moorings, a lobsterman and his helper returning from a day of fishing, and several small skiffs that lobstermen use to get to and from their boats. In the background is the lobster shack that is commonly known as 'Motif #1.' It is one of the most gazed-upon, sketched, painted, and photographed subjects along the entire New England coastline.

PREFACE

As a young lad growing up in the city of Beverly, Massachusetts, I lived in very close proximity to the waterfront, the wharf, and the long line of piers that extended along Water Street in that city which is located about halfway between the cities of Boston and Gloucester.

I spent those early years of my life "just hanging around" the waterfront and at the Beverly Fish and Lobster Company where I helped to fillet groundfish that was trucked in from the bustling fishing port of Gloucester. On many weekends I sold fresh fish, crabs, and live lobsters to the regulars who would buy their seafood and shellfish only from the local fish market. Live "Chicken" lobsters at that time were selling for 69 cents a pound! I used to look forward to riding along in the bait truck that would travel to Gloucester to be loaded up with redfish for the local lobstermen. I didn't, however, relish with any degree of enthusiasm the wet, messy, and smelly job of forking those redfish into the lobstermen's bait barrels that were lined up and stacked deep along the fish wharf.

Quite often one of the local lobstermen would take me along fishing with him, and being somewhat of an inquisitive young boy, I always had a lot of questions to ask as he was hauling his traps. I was always excited about those live and kicking lobsters that the lobsterman would bring aboard his boat, but even back then I looked upon the job of a lobsterman as just being too much work. In those carefree days of my youth, "lobstering" was simply too much huffing, puffing, and grunting to suit me, so I decided right then and there that my life's work would not be that of a lobsterman.

I spent so much time on the waterfront, including skipping school to do so, that I was often referred to as a "wharf rat." There were always people who would be on the lookout for Mr. Roundy, the truant officer, who usually had a good idea where he could find me when I did skip school. And although there was one occasion where I almost got caught hiding in an empty mackerel barrel in the cooler, never once did "Mr. Round-up Roundy" ever find me in all of his searches. My dear mother often remarked that "she could smell Robby coming up Silver Court" when I did manage to pull myself away from the fishy-smelling odors that permeated Beverly's inner harbor.

Over the years I have spent many pleasurable hours "going out" with lobster fishermen in Maine and Massachusetts. I liked to pick up little tidbits of knowledge about the craft of lobstering by chatting with these seafarers as they were hauling, moving, and setting their traps. It was in this environment that I could observe at close hand the methods and techniques being used in the fishing of their traps. Similarly, countless hours have been expended over the years in chatting with lobster marine scientists and university people so that I might gain a better and more accurate insight about lobsters from these people who dedicate their life's work to this pursuit.

A most significant researcher of the American lobster was Francis H. Herrick (1858-1940) who in 1903 began to work on his classic book, the *"Natural History of the American Lobster."* This very detailed study of the American Lobster was later published in 1909 as a Bulletin of the Bureau of Fisheries, and in my opinion, constitutes the only really comprehensive study of the American lobster that was known at that time. While researching Herrick's masterpiece, I found it incredible that the most major concerns about the lobster fishery at the beginning of the 20th Century still lingered and remained as such at the beginning of the 21st Century.

Francis H. Herrick's study of the lobster is a monumental piece of research that includes in finite detail every aspect of the lobster's life cycle and the dominant practices and conditions that have an impact on the American lobster fishery. And from the profusion of research material that has been at my disposal, there has probably been no other lobster marine scientist who has ever assembled under one cover a more thorough study of the American lobster than Francis H. Herrick.

In addition to the work of Herrick, much of the research material that has been utilized in the writing of this book has been derived from the research of lobster marine scientists from the two major lobster-producing states of Maine and Massachusetts The use of only these sources is not intended to be a slight to the other nine lobster-producing states of New Hampshire, Rhode Island, Connecticut, New York, New Jersey, Maryland, Delaware, Virginia, and North Carolina. Nor have I overlooked the Canadian lobster fishery, which is, in fact, the largest lobster-producing entity of the American lobster. The data from Maine and Massachusetts, and especially from Maine, has been relied upon with regard to a wide range of topics such as the lobster's life cycle, conservation measures, fishing effort, catch yield, migration studies, mortality rates, and so forth. To introduce very similar data, if it was indeed available at all, from the other lobster-producing states and Canada would only serve to confound the reader, which is certainly not the intention of this writing.

Throughout the pages of this book the reader will notice that a great deal of attention has been directed to the role of the lobster fisherman. And rightfully so! "The Lobster Fisherman" which appears in Chapter 5 is my personal tribute to those rugged individuals who fish for the lobster in good times and bad and often under intolerable weather conditions. In this regard, I would like to extend a personal word of thanks to those lobster fishermen who have allowed me on their boats and have chatted with me at dockside.

My sincere thanks are also extended to the many lobster marine scientists who have provided essential research material, all of which has greatly aided me in the writing of this book. I would especially like to thank Jay S. Krouse, former lobster marine scientist at the Maine Department of Research Resources Laboratory at West Boothbay Harbor. In his capacity as the lead researcher of lobsters and crabs for the Maine DMR, Mr. Krouse has been extremely helpful to me regarding the story that I wished to tell about this remarkable crustacean that defies extinction.

One major theme that ripples here and there throughout the pages of this book is that of a fishery that was targeted to face depletion and even outright destruction. While the lobster fishery has appeared to have been depleted at times over the years, the sinister predictions about the fishery's outright destruction have shown to be just predictions that have been proved to be invalid. Despite the warnings of Herrick and others, the catch of the American lobster remains bountiful, and this is essentially so during the most recent years of the 20th Century. Had it not been for the alarm and concern of Herrick and other researchers that followed him, perhaps the lobster fishery would have become depleted, and perhaps the fishery would have become destroyed. But the lobster did not perish, and as we shall come to understand, the lobster continues to flourish into the 21st Century.

It is important to note that certain words and expressions presented throughout this book have been printed in italics and sometimes in italics and/or quotation marks. These words and expressions are often nautical in nature, or pertain to the lobster fishery in general, and have been set forth for that reason. Many of these words and expressions can be found in the chapter entitled "A Dictionary Of Important Lobster And Lobstering Terms."

A conscious effort has been made in the writing of this book to break down the subject matter in such a manner that it can be quickly and easily comprehended by readers of most any age group. In this regard, it is my sincere hope that anyone who has an interest in the American lobster will take as much pleasure in reading "The Tale of the Lobster" as I have had in writing it.

Robert Delano Martin
Beverly, Massachusetts

Adult female American red lobster (top) and adult male American lobster (bottom). Drawn from life. From Francis H. Herrick, *"Natural History of the American Lobster."* It should be noted that the top specimen is not a drawing of a lobster "out-of-the-pot," but a red pigmented lobster only rarely caught in its natural environment or scientifically natured in a laboratory environment.

Chapter 1. The American Lobster – A Threatened Marine Resource?

The American lobster, Homarus americanus, has attracted the attention of people from all walks of life throughout the world. H. americanus is probably the most easily identifiable specie of crustaceans that inhabit the bottom of the ocean along the northeastern coastal range of North America.

Perhaps our fascination with the lobster is in the way it behaves in a lobster pool; or perhaps it is the beautiful color pigmentation of the hard shell that encases its body. Possibly our attraction to the lobster is heightened by observing it alive, kicking and moving about, when just a few hours or days earlier it was in its natural habitat, out of captivity, and meandering here and there on the bottom of the ocean. Whatever the attention or attraction, the American lobster will be examined and scrutinized in the pages that follow. In this regard, the lobster will receive its just due!

The range of the American lobster extends from off the coast of Newfoundland in the north to off the coast of North Carolina in the south. The lobster is caught close to shore (inshore lobstering) and far out into the ocean (offshore lobstering). The crustacean lives in water depths as little as two fathoms (12 feet) to as much as fifty fathoms (300 feet). Lobsters are caught in traps, sometimes called "pots," and are often pulled up from the bottom in the nets of dragger fishermen, by long-line fishermen, and by SCUBA divers who come across them while engaging in that recreational endeavor. Many lobsters are taken out of their natural habitat as a direct result of severe coastal storms and by strong tidal currents. On occasion, lobsters are washed ashore in traps that are smashed and driven up on the often-rocky coastline by violent storms. Historical records tell us that the native Indians taught the early settlers how to catch and use lobsters. On the North American continent, lobsters were caught for the first time during the early part of the 17th Century by Pilgrims and Englishmen who made up the Massachusetts Bay Colony. The abundance and the large size of these lobsters were commonplace for these earlier settlers. The Minister Higginson, writing of Salem (Massachusetts) lobsters, said that "many weighed 25 pounds apiece, and that the least boy in the plantation may catch and eat what he will of them."

The early settlers of New England did not look upon the lobster as a delicacy. Lobsters were incredibly abundant and were easily caught by using simple devices such as hook and line, nets, and gaffs. At low tide, lobsters could easily be plucked out from under rocks and seaweed. With just a little probing around with a stick or gaff, they were often found hiding in rock crevices that made up a good portion of the New England and Canadian Maritime Provinces. Indeed, lobsters were so plentiful during the 17th Century that they were often turned into the soil as a fertilizer for corn and other vegetables.

Recognizing this incredible abundance of lobsters, and with just a little stretch of the imagination, one might now turn back the clock of time hundreds of years earlier and visualize an ocean bottom that was teeming with lobsters. There were no lobster traps or lobster fishermen as we know of them today. Our visualization portrays an ocean bottom off the coast of New England and Canada as being literally black with lobsters; a bottom so profuse with lobsters of all sizes, that there must have been very little space for other marine life to coexist with them there. They were left unhampered in the deep confines of the ocean to molt, mate, and produce in their own numbers. It was there in that dark and gloomy environment that these legions of lobsters pitched a day-to-day battle of survival, not only with their own numbers, but also with other species of marine life that had to subsist with them in the same ocean- bottom environment.

The commercial lobster industry did not show any real signs of accelerated development until about the beginning of the 19th Century. The Cape Cod region of Massachusetts then started to attract fishermen and their *smacks* from Connecticut and other nearby New England states. This new breed of the commercial lobster fishery furnished most of the supply of lobsters that were shipped to the large metropolitan cities of Boston and New York. But, for some unexplainable reason, the commercial lobster industry was not introduced and extended into the State of Maine until about the Year 1840. It was at this time that commercial fishermen entered the waters off the coast of Maine and into the area known as Casco Bay.

An initial concern about the over-fishing of the lobster, as well as for the protection of the fishermen themselves, dates back to the Year 1812. Francis H. Herrick talks about a Dr. Rathbone, whom he quotes as saying:

> "In 1812, the citizens of Provincetown, realizing the dangers of exhausting their fishing grounds, succeeded in having a protective law enacted through the state legislature, apparently the first but not the last of its kind. For legal restrictions, including this statute, have been in force ever since. But this measure was designed to protect the fishermen rather than the lobster, for it was merely declared illegal for anyone not a resident of the Commonwealth to take lobsters from Provincetown without a permit. The laws later

enacted proved of little or no avail. By 1880 the period of prosperity had long passed, and few lobsters were then taken From the Cape…

"This great fishery was thus rapidly exhausted by over-fishing and it has never recuperated."

And in 1895, Herrick sounded an ominous warning concerning the possible depletion of the American lobster fishery along its entire range:

"Civilized man is sweeping off the face of the earth one after the other some of its most interesting and valuable animals, by lack for foresight and selfish zeal unworthy of the savage… The ocean indeed seems to be inexhaustible in its animal life as it apparently is in extent and fathomless depth, but we are apt to forget that marine animals may be as restricted in their distribution as terrestrial forms, and as nicely adapted to their environment… And the American lobster occupies only a narrow strip along a part of the North Atlantic coast, and while it is not possible to exterminate such an animal, it is possible to so reduce its numbers that its fishing becomes unprofitable, as has already been done in some places.

"The only ways open to secure an increase in the lobster are to protect the spawn lobsters, or to take the eggs from the lobsters themselves and hatch them artificially."

In a booklet entitled, "Harvesters of the Sea," the rise and fall of the Maine lobster fishery is clearly depicted:

"The first pack of sealed goods of any kind to be put up in this country was lobster; and quite appropriately this was accomplished in 1843 at Eastport, the town which so often led the way in the early days of seafood processing. A luxury item which commanded a high price, canned lobster looked like a potential goldmine. Unfortunately, the lobster supply, at first believed unlimited and inexhaustible, proved to be neither.

"Wide-spread exploitation of this fishery completely revolutionized Maine's commercial fisheries and contributed largely to the popularity of the state's seafood products. But it also caused the State Legislature to pass a series of restrictive laws designed to prevent the total destruction of the inshore lobster population.

"Though fishing technology might be said to have advanced in a straight line, individual fisheries sometimes waxed and waned.

"The lobster is one example – booming as the canning industry boomed, collapsing as the supply became short. Then gradually recovering as conservation measures were put into effect."

It was researcher Francis H. Herrick who once again sounded the warning bell over his concern about a possible depletion of the lobster resource. His concern about the egg supply of the American lobster was one of his chief concerns, as he related in his book, the *"The Natural History of the American Lobster"*:

"The problem before us is how to aid nature in restoring and maintaining an equilibrium of numbers of the specie, or how to increase the number of adult animals raised from the eggs. It concerns not only the fishermen who earn a livelihood through the fishery, or the dealer who has capital at stake, but the public of many lands, in fact, everyone in the Western Hemisphere at least who likes the lobster for food. When the decline of the already depleted fisheries became a serious menace, protection was sought in legislation, but since the lobster supply of this country is drawn from many states and the Maritime Provinces as well, no uniformity of laws or methods was to be expected. Each state enacted its own laws, which were often used at variance, unscientific, and subject to continuous change. Up to the present time every effort to check the constant and ever-increasing drain on this fishery has signally failed, which shows that either the laws are defective or that the means of enforcing them are inefficient.

"A sound and essentially uniform code of laws for the entire fishery is plainly demanded if legal restrictions are to be of much avail."

The ongoing concern about the decline and diminishing abundance of the lobster is not totally confined to the United States and Canadian lobster fisheries.

Robert Stewart, a lobster fisherman from Scotland, dramatizes the former abundance of lobsters and their subsequent decline:

> "There are many folk around here (Moray Firth area of Scotland) who can remember childhood days spent in crab and lobster hunting along the shoreline. Provided with the large type of fishbasket used on boats, they would go down to the water and find large lobsters and crabs in profusion in the rock pools and under clumps of seaweed. They would often return with their baskets brimming over. But now the story is different...
> "Fifty or sixty years ago it was possible to make big catches from the beach, but then the catchers took only enough for their needs or for their neighbors. One could find many lobsters in each hole and crabs hiding under each clump of seaweed, but over the years there has been a steady decline in the stocks for both lobster and crab."

Kendall Merriam in his book, "The Illustrated Dictionary of Lobstering," draws still another illustration of the frustrating decline of the lobster fishery in the State of Maine:

> "One day some summer *soon*, a lobsterman will come back halfway through hauling, frustrated and enraged. He will say, 'There was no sense hauling. I only got two lobsters in fifty traps.'
> "That was a far cry from the summer of 1952, when my brother, Paul, hauled up a trap in the cove at Matinic which had eighty-two lobsters in it, fourteen of them *counters*. And that was a long way from when men picked them up from under the seaweed, Gaffed them in the shallows, and caught a dozen of them in fifteen minutes on hoop Nets. Lobsters used to wash ashore in windrows after the big storms – but that hasn't happened in decades."

And Dr. John T. Hughes, former Director of the Massachusetts State Lobster Hatchery and Research Station at Martha's Vineyard, communicated to the author that *"LOBSTERS ARE AN ENDANGERED SPECIES!"* This candid assessment of the lobster fishery was made in 1978, and it was advanced from a marine lobster biologist and scientist who is recognized as an authority on the American lobster.

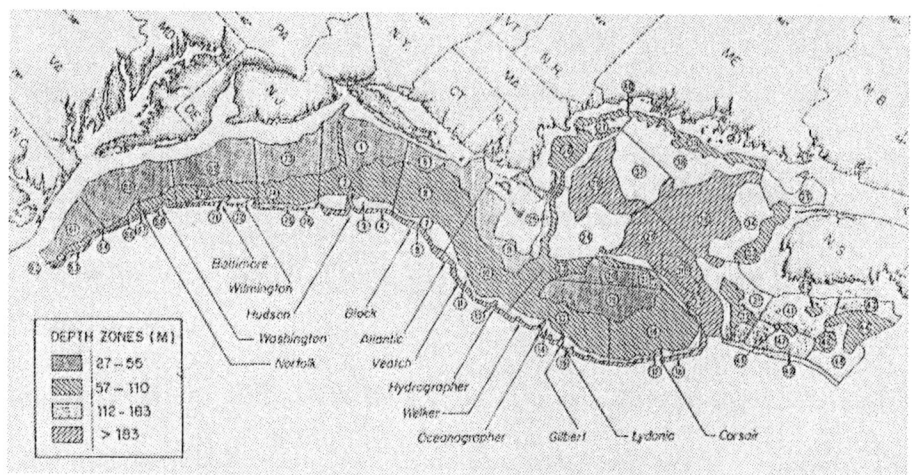

Location of offshore submarine canyons on the outer continental shelf and sampling strata for research vessel surveys.

A virtual mountain of data concerning the lobster more than supports the fact that the animal is being both intensively and extensively fished along its entire range by fishermen working both the inshore and offshore lobster habitats. Fishing close to the shoreline, the inshore lobstermen account for approximately 80% of the total domestic production of lobsters. The Canadian and United States inshore fishery, as we have seen, extends along a narrow swath of coastline from the Strait of Belle Isle, Newfoundland in the north to Cape Hatteras, North Carolina in the south, but the greatest abundance of lobsters is concentrated between Nova Scotia and New York. Lobstermen working this inshore habitat customarily set their traps in water depths ranging from about 5 fathoms (30 feet) to 15 fathoms (90 feet).

When lobster traps are set extremely close to islands, ledge croppings, and the like, the distance from the surface to the bottom is so small that in clear and un-muddied water one can actually see the traps laying at rest on the bottom.

Trap gear used in trapping lobsters in the offshore lobster habitats, however, require that they often be set in hundreds of feet of water because of the deepness of the ocean. The offshore fishing areas, referred to as *stratas* by the National Ocean and Atmospheric Administration (NOAA), extend along the snake-like Continental Shelf that extends from Georges Bank and the Gulf of Maine southward to North Carolina. The Continental Shelf is broken at intervals by a total of fourteen "*canyons.*" Beyond the southern-most point off North Carolina, the American lobster ceases to exist. This point southward then becomes the habitat of the clawless spiny lobster, "*Palinurus.*"

Offshore trawl surveys conducted on a regular basis by NOAA indicate that in the spring there are larger concentrations of lobsters along the edge of the Continental Shelf and in the Gulf of Maine. These surveys also reveal that there is a much wider dispersion and scattered distribution of lobsters in the more shallow waters off the Scotian Shelf, Georges Bank, and Southern New England. This difference in the distribution of the lobster population, especially in the spring, is essentially attributed to the migratory movement of the larger offshore lobsters. According to marine biologists, a significant number of the offshore female lobsters migrate shoalward to warmer waters in the spring for the purpose of hatching their eggs, to molt, and to mate with male lobsters. These migrating lobsters, unless trapped in the meantime, most often return to the deeper offshore waters in the autumn as the coastal waters begin to cool.

The American lobster comprises one of the most valuable and important commercial fisheries in the northeastern section of the United States. Compared to other types of marine life, the American lobster is more economically important than even the "mighty cod," the haddock, the herring, and other dwellers that live in this salt water habitat.

On the basis of data derived from commercial landings, port and ocean sampling, and state and federal *"trawl"* surveys, it is estimated that approximately 80% of the landings of the American lobster are from the *"inshore"* trap fishery and in state waters within 3 miles from the coast. The lobster *"trap"* constitutes the primary device used to catch the lobster (98%), while *"trawls"* dragged along the ocean floor account for the other 2% of the landings.

Intensive fishing *"effort"* has motivated and enticed some lobster fishermen to go farther out into the ocean in source of their prey in the deep water habitat where lobsters abound in great number and in greater size. Significant deep-water habitats are generally located in federal waters that extend from 3 to 20 miles from the coastline and in ocean areas traditionally known as the far-out reaches of the Gulf of Maine and the Georges Bank. The American lobster is considerably abundant in and around the Continental Shelf and its *"canyons"* that stretch for hundreds of miles off the coast of the northeastern and middle Atlantic states.

Virtually all of the United States American lobster fishery exists in state and federal waters from three basic *"Statistical Stock Areas"* that have been classified by the National Marine Fisheries Service as the Gulf of Maine (GOM), the Georges Bank and the southern New England outer shelf (GBS), and the area classified as South of Cape Cod to Long Island Sound SCCLIS). These three stock areas, their relative size, and their contribution to the American lobster fishery, are summarized as follows:

Stock Area	**Area Size**	**Area Size**
GBS	Largest Stock Area	3rd Largest In Landings
GOM	2nd Largest Stock Area	1st In Landings
SCCLIS	Smallest Stock Area	2nd In Landings

While the overwhelming percentage of American lobster landings are derived from the inshore lobster fishery, it is important to note that in 1998 two of the eleven lobster-producing states, Maine and Massachusetts, accounted for approximately 75% of the total landings, and four of the eleven southern-most lobster-producing states, Delaware, Maryland, Virginia, and North Carolina, contributed only a smidgeon of the total landings. Reference is made to Chapter 6 for more details regarding *"catch"* (landings) for the seven major lobster-producing states.

Has the American lobster marine specie been threatened? Yes! Has this marine animal been destroyed? No! Has the lobster habitat been depleted? No! In spite of intensive and extensive fishing pressure by lobster harvesters, there have been conservation measures put into place over the years to protect the animal from depletion and outright destruction.

What is the status of the American lobster fishery in United States waters in the year 2001? Ask the lobstermen and there will be one set of answers - mostly with the twist that, "Hey, the annual catch has been increasing; 1998 was a great year, and year 1999 was a banner year for us. What's all the fuss about? We've done more than our share of putting lobster conservation measures into place and in cooperating with the governmental Powers-To-Be; we're being over-regulated and we don't like it one single bit. Just leave us alone and everything will be all right for us and the lobster fishery!"

Flip the coin and there is an echo in the wilderness - the state and federal governmental bodies that exist to control, safeguard, and manage the American lobster fishery. The Atlantic States Marine Fisheries Commission in its voluminous March 2000 *"American Lobster Stock Assessment & Peer Review Summary"* stated specific concerns and possible future resource management options, a few of which have been excerpted as follows: And make no mistake about it, the state and federal governmental bodies, and especially the federal bodies that regulate to a large degree what happens to the fisheries industry, have a lot to say about what has happened and what might or could happen to any aspect of the marine fisheries, from the cod, the haddock, the bass, the herring, the lobster, and virtually every other form of marine life that suggests even the slightest chance of being threatened. The people and money resources targeted to carry on research and resource management of the marine fisheries have been made available in the past, and these resources are and well- entrenched to carry on such pursuits in the future. They are, so to speak, the *"regulators of the deep,"* and although all commercial fishermen, lobster fishermen and otherwise, may not totally agree with all of their postulations, these state and federal marine resource managers are a voice to be heard - and it is probably apparent to anyone who has delved into marine resource management that they have much to offer, are well intentioned, and are often correct in their assumptions and predictions regarding the present and future status of virtually every form of marine life that makes its habitat in ocean waters. The matter of *"resource management"* is discussed to some degree in Chapter 6, *"Practices and Conditions That Influence the Lobster Fishery,"* and while it often appears to be a matter of *"We say Vs. They Say,"* there is provided in the following some key excerpts selected from the March 2000 document published by the Atlantic States Marine Fisheries Commission in its *"American Lobster Stock Assessment and Peer Review"* report. Key phrases in the following excerpts have been set forth *"within quotation marks"* for the purpose of putting emphasis on the role of the federal and state government in the regulation of the American lobster fishery in the United States:

> "The fishery in state waters is managed through the Atlantic States Marine Fisheries Commission's (ASMFC) American Lobster Management Board. The Board developed Amendment 3 to the Interstate Fishery Management Plan for American Lobster in December 1997, "which is implemented through state regulation." The plan, when fully implemented, "is designed to minimize the chance of population collapse" due to recruitment failure. "The goal of Amendment 3 is to have a healthy American lobster resource and management regime, which provides for sustained harvest, maintains appropriate opportunities for participation, and provides for cooperative development of conservation measures by all stakeholders."

"Amendment 3 outlines a rebuilding schedule to end overfishing by 2005. The primary management resources include a minimum size, protection of egg-bearing females, and trap limits. The Board also recommends that the Secretary of Commerce implement consistent management measures in adjacent federal waters. The intent is to establish management measures that are compatible with the implementation of Amendment 3, end overfishing, and rebuild the stocks."

"The most recent American lobster stock assessment was conducted by the 22nd Stock Assessment Review Committee (SARC) in 1996 (NEFSC 1996). According to the last assessment, the 22nd SARC 'warned that when stock collapse has occurred in other lobster and crustacean stocks, it has been sudden, and stock rebuilding has required decades.' 'It advised that fishing mortality be substantially reduced in the U.S. lobster fishery to decrease the possibility of stock collapse.'"

"A new assessment was commissioned in 1998 to provide up-to-date information on stock status for management purposes." The ASMFC Lobster Technical Committee was therefore charged with forming a Lobster Stock Assessment Sub-Committee (LSASC) to carry out this task. The Sub-Committee was formed in January 1999 and carried out this assignment during the year, subject to the following terms of reference:

1. Compile data needed for stock assessment purposes, updating databases to include most recent information available.
2. Retain the three stock assessment areas and for each area:
    - Estimate current levels of egg production, abundance and mortality rates;
    - Evaluate uncertainty associated with stock status indices;
    - Evaluate historical trends in population abundance, fishing mortality and recruitment, using population dynamics models and other indices;
    - Review and up-date biological reference points used to evaluate stock status.
3. Develop analyses that could explain why the abundance and recruitment of lobsters has continued to increase in spite of the overfished status of the resource.
4. Review reports made in 1996 by the Lobster Peer Review Panel and Stock Assessment Review Committee, evaluate the current status of each recommendation, and act on any remaining recommendations, which the Assessment Sub-Committee believes, are appropriate

and useful for resource assessment purposes, and for which there is sufficient time.
5. Evaluate any other appropriate stock assessment methods and approaches that the Assessment Sub-Committee believes are needed and has time to develop and prepare analyses for review by the ASMFC Peer Review Panel."

Consensus
"The LSASC (Lobster Stock Assessment Sub-Committee) strove to reach consensus on data, methods and results in this stock assessment. Most of the conclusions in this assessment are consensus opinions. This assessment endeavors to present the full range of technical uncertainty about the status of the lobster stocks."

Having reviewed the aspects of state and federal and state governmental research, findings, opinions, conjectures, and recommendations regarding the United States lobster fishery, and the alleged *"threatening and possible depletion"* of this valuable marine resource, it is important that we proceed to make a brief inspection of the animal itself, the hardy souls that fish for it, and other considerations that make the American lobster such a sought-after delicacy throughout the world.

Chapter 2. The Anatomy And Biology Of The American Lobster

The lobster is a marine invertebrate, an animal without a backbone or spine. Since lobsters have an external skeleton, they are classified as arthropods. And because lobsters possess hard shell casings to protect their bodies, they are known as crustaceans; and since they have ten legs, they are further classified as decapods. Thus, the phylum is Arthropoda, the class is Crustacea, and the order is Decapoda.

Look into a lobster tank at a local fish market or supermarket and what do you see? Lobsters with two claws, lobsters with only one claw, and many of them crawling over one another in an attempt to gain access to the darkest-most recesses of the tank. You would see lobsters of various sizes and weight, ranging from the smaller size *chicken* lobsters to the larger size *select* and *jumbo* lobsters. Some will have large claws of approximately the same size, but most will have one large claw that is either smaller or larger than the other. There might be a lobster with only one large claw. That lobster is known as a *cull* lobster. A lobster with a deformed claw might also be seen, and it is in this condition because the claw was not perfectly reformed during the building of a new shell following a *molt*, or shedding of the old shell.

Peering into a *holding tank* at a lobster hatchery and research station would reveal all of these differences in lobsters – and even much more. For example, a visit to the Massachusetts Lobster Hatchery and Research Station at Martha's Vineyard might afford one the experience of viewing a red lobster, a blue lobster, a spotted calico lobster, and possibly a lobster that is one-half jet black and one-half fire engine red! To be able to view such a lobster would be a real special occasion, because such an animal is rarely found or recovered in nature.

The marine biologist or an assistant working with lobsters at the research station might even be coaxed into holding up a *berried* female lobster with eggs attached to the under part of the tail section. The eggs of the female lobster could range in color from dark to golden brown, the darker brown eggs indicating immature eggs, and the golden brown eggs indicating mature eggs that any day might be ready to hatch. When this hatching does take place, the thousands of minuscule *fry* are kept constantly whirling about in the water of a circular *rearing tank* until they will have grown to the proper life cycle stage when they can be released into the ocean.

THE SENSORY FUNCTIONS

Sight

Despite the lobster having compound eyes consisting of some 10,000 facets, it is generally believed that the animal is of rather dull sight. A lobster in captivity has a marked tendency to shun light by crawling into the darkest recesses of a holding tank. It is judged that the lobster's sight is incapable of defining a clear image; its eyes can only see a blurred image at best, and reacts to changes in the intensity of the light and to anything in motion. This can best be demonstrated by causing a shadow, as by a hand, to pass in front of the lobster's eyes. The large crusher and ripper claws will usually swing wide open and the lobster will attempt to attack the shadow. At the same time, the several appendages that make up the various mouthparts will be set in motion, and the first and second pair of walking legs will begin to become very active. But it is important to understand that the lobster is only reacting to the shadow cast by the hand, and that the lobster probably sees the shadow as only a blurred and undefined image.

Hearing

A lobster does not have ears in the true sense of the word. It is believed that the lobster "hears" only the vibrations set forth in motion by some force or activity in the lobster's immediate surroundings. The vibrations are detected and picked up by the lobster's sensory organs and by the shell of the lobster itself. In this regard, it is thought that severe vibrations and disturbances such as thunder storms, fireworks displays, and even the nets of commercial fishermen dragging their nets along the ocean floor, will be quickly detected by the lobster and send it into hiding.

Smelling And Tasting

Not only is a lobster without ears to hear, but the animal is also without a nose to smell. For smelling and tasting, as well as for touching, the lobster is greatly dependent upon the profusion of thousands of delicate sensory hairs located throughout the body. These sensory hairs are called *satae*, and are located on the lobster's main parts and appendages. Close scrutiny of a lobster will reveal these sensory hairs in many other areas [refer to the illustration that accompany this chapter for an easy reference to these parts and appendages]: on the *ripper claw*, on certain surfaces of the *attenules*, on the various appendages that make up the mouth parts, and on several appendages of the four pairs of *walking legs*. These sensory hairs can also be found along the very edges of the shell of the *carapace* and on the hard shell that encases the tail section, the *abdomen* of the lobster. A very close examination of the lobster will also reveal an abundance of sensory hairs at the edges of the flat blade-like appendages that make up the *tailfan* of the lobster. Flipping a lobster over onto its backside will show evidence of a profusion of delicate sensory hairs on the *pleopods* of the lobster. The pleopods are the lobster's abdominal appendages that are commonly referred to as *swimmerets*.

These sensory hairs, some of which also serve as organs of touch, are essential organs of smell and taste in that they serve as chemical receptors that quickly pick up the presence of *pheremones* and food in the lobster's immediate environment on the floor of the ocean.

Touching

Some of the sensory hairs that abound in the lobster are used as organs of touch. These are especially abundant on the dactylus, *propodus*, and *carpus* of the first two pairs of walking legs; also on the propodus and carpus of the last two pairs of walking legs, as well as in and around the teeth-like structure of the long and slender ripper claw. The long branched *second antenna* and *flagellum* are very important touch organs and are often referred to as the lobster's *feelers* as it moves about on the bottom.

The Major Body Parts And Their Functions

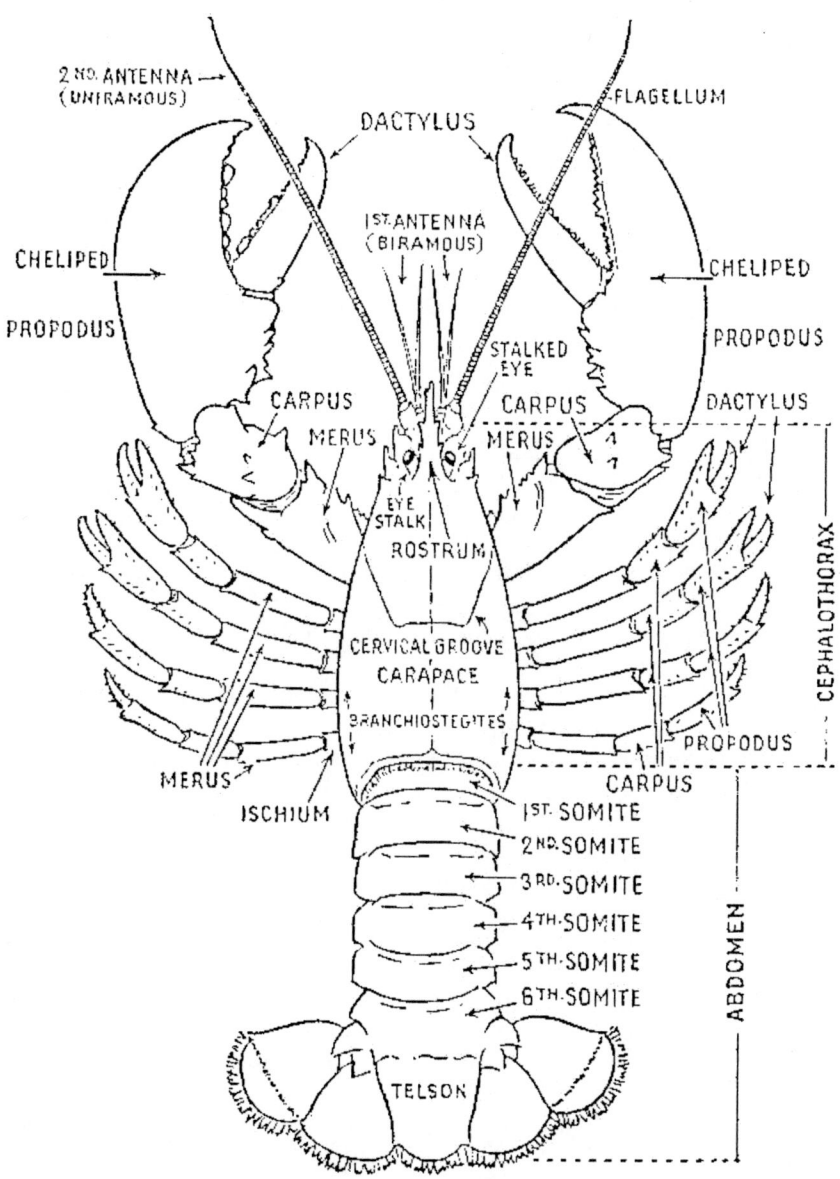

This illustration shows some of the major body parts of the American lobster. The "ripper claw" is shown on the right and the "crusher claw" is on the left. However, these claws are often reversed and there are instances when a lobster will have either two ripper claws or two crusher claws.

Because the lobster is a complex marine invertebrate, a careful examination of the animal should be made in order to fully comprehend and further appreciate the complexity of this forester of the deep. Referring to the illustrations provided in this chapter will enhance this understanding. If possible, the going out and the buying of a live lobster would certainly be of benefit to anyone who wishes to see at first-hand the body parts of the lobster and the functions that they perform.

The Protective Shell Casing

As one looks down upon the lobster, most of the lobster's hard *exoskeleton* comes into view. This outer skeleton is made up of two main sections, the *cephalothorax*, which is essentially the head and body of the lobster, and the *abdomen*, which most people know as the tail or tail section. The shell of the lobster is comprised of a three-layered structure known as *chitin*, which forms a solid crust over the *cuticle* that lies directly beneath. This hard shell consists of a concentration of lime salts and other minerals that are absorbed by the lobster both before and after a molt takes place. Where the two sections of the lobster join together, the shell structure is noticeably softer, thinner, and transparent. The increased flexibility in this dividing point in the lobster enables it to bend, flex, and move about freely. The softness of the shell structure is especially apparent at the juncture where the shell of the carapace meets with the shell of the abdomen, in the knuckle appendages of the large claws, and on each of the six *somites* on the underside of the abdomen. It is interesting to note, as we shall later see, that the hard shell of the carapace is the only part of the lobster's shell that is *broken* when the lobster molts in order to cast off its old shell that will be replaced with a new one.

The Carapace, Rostrum, and Eyestalks

A lobster is best handled by grasping it by the carapace, which is the *back shell*, or *dorsal shield* of the lobster. The shell of the carapace encases and protects the vast majority of the lobster's vital organs. Some of these organs, though not all, include the heart, brain, gills, stomach, and liver. At the anterior end of the carapace, and centered between the two *stalked eyes*, is the jutting and sharply pointed *rostrum*. The rostrum features nine or so sharp and pointed spines that are thought to be a means of protection of the lobster's eyes, the latter of which are deeply set in eye sockets on each side of the rostrum. For the lobster fisherman, the eye sockets are a very important reference point because they are used when measuring a lobster to determine whether or not it is of legal size. Lobster marine biologists also us the eye sockets to measure from when conducting research studies to determine a lobster's overall carapace length.

The Great Chelipeds

The American lobster sports five pairs of *thoracic walking legs*, the first pair of which are classified as the *great chelipeds* (pronounced kee-lo-peds). Most people refer to them as the *large claws*, or more simple, *the claws*. The larger and more bulky of the two is called the *crusher claw*; the longer and more slender is commonly referred to as the *ripper claw*, or *pincer claw*. While these two large claws are biologically classified as legs, they actually serve little or no ambulatory functions at all. Rather, they function to protect the lobster when being attacked, and are used as the "hands" and "fingers" when seizing, holding, crushing, and ripping food apart before it is passed back to the various mouth parts for further processing.

The functioning of the lobster's large claws is directly observable in a controlled environment, such as an aquarium or lobster tank. The crusher claw will go to work to break down shell formations or other sources of food for the ripper claw to deal with. The ripper claw is much quicker to work on food, such as a scrap of shrimp or a mussel, as compared to the larger crusher claw, and it will take this claw virtually no time at all to shred and rip the food apart.

The claws of a mature male lobster are generally larger and heavier that those of a mature female lobster of the same size. When the lobster is on the move, it will have the tendency to keep its claws straightforward and held close to the body. A lobster that is threatened will open its claws as far apart as possible, in a spread-eagle manner, as if in an attitude of attack.

In the jargon of the lobsterman, a lobster that carries its crusher claw on the right side is called a "right-handed lobster," while a lobster that carries its crusher claw on the left side is called a "left-handed lobster." However, there have been occasions, although relatively few in number, when a lobster has either two crusher claws or two ripper claws.

Many people who buy live lobsters are scared-to-death of these large claws! But even lobstermen handle their prey with caution and respect when removing them from the trap. To prevent injury by being "snipped" by either of the large claws, lobstermen make it a practice to shut them tight with heavy rubber bands. Rubber-banding of the large claws also minimizes injury to other people along the distribution line, including the distant-end consumers who have purchased their lobsters alive from a retail outlet and then have to deal with these live and often scrappy lobsters when they get them home. The lobster fisherman, furthermore, has a parochial reason for banding the large claws: if he is to hold onto his catch for an extended period of time, and without a proper supply of food, his lobsters – without banding could become cannibalistic.

This condition might very well lead to injury, disease, and ultimate death to other lobsters, and a lobsterman is not about to allow this to happen.

The Knuckle Appendages

In between the *propodus* of each of the large claws are the *carpus* and *merus* appendages. In addition to functioning as jointed limbs, these appendages, along with the propodus itself, have the capability of being "thrown off" by a lobster that is seized by a predator. When the lobster is thus threatened, it has the ability to sacrifice either of the great chelipeds as an escape mechanism to save its life. This is known as *defensive mutilation, autotomy,* or *self-amputation.* The claw is broken off at a specific place, called the "breaking plane," near its base. Fortunately, however, the large chelipeds have the capability of being regenerated However, in relation to the breaking plane, there has been discovered an interesting *interlocking mechanism,* an adjustment by the lobster which makes it virtually impossible for an adversary to pull or twist off a large cheliped of the lobster. So it might be theorized that if a lobster feels it can survive an encounter with an adversary that has a hold on one or both of its large chelipeds, it will bring into action the interlocking mechanism, and "hold on for dear life!" If not, it will throw the claw, or claws, in order to loosen the hold, retreat, and escape from its combatant.

The knuckle appendages that make up the walking legs of the lobster are referred to as the *propodus*, the *carpus*, the *merus*, and the *ischium*. These slender walking legs are often lost while held in storage in floating *lobster cars* where mutilation takes place. According to Herrick, "Many, if not all, of the appendages, when mutilated or removed, are capable of regeneration, the time required for the process depending upon the proximity of the succeeding molt, the vigor of the animal, and the temperature of the water."

The Antennules

The *antennules of* the lobster, also known as the *first Antenna (Biramous)* is composed of two pairs of branched antennae that measure about two inches in length in a 1½ pound lobster. The edges of the outer sides contain sensory hairs that are used for the detection of pheremones, food, chemicals, and other odors in the water. It can be observed that the four blade-like appendages that comprise the antennules will go into a constant frenzied motion when food is placed anywhere close to the lobster. However, it is believed that a lobster that has had its first antenna severed will find food only if it just happens to stumble upon it!

The Whipped Antenna

The antennas of the lobster are comprised of one pair of long, slender, and tubular whipped antennae. One of the two antennas is referred to as the second *Antenna (Uniramous)*, with the other branch being classified the *flagellum*. These two jointed antennas measure approximately eight inches in a 1½ pound lobster. The lobster, when moving about, or when in a deliberate search for food, will extend these branched appendages directly out in front as they "sweep" the water in the immediate environment. When threatened, the branched antenna will usually be swept back and carried close to the body. While these *feelers* of the lobster are thought to be used primarily as chemical detectors, they are also used for tactical purposes as organs of touch. The lobster, therefore, uses these antennas for detecting the presence of food and other stimuli, for detection of disturbances in the water, and for touch-testing objects on the ocean floor.

An analysis of lobsters found in any lobsterman's daily catch reveals a large number of lobsters that have mutilated whipped antennas. Some of these lobsters have had part of the antennas snipped off, while other lobsters might have no whipped antennas at all. This suggests that there is a lot of scrapping going on "down there," and the antennas are very vulnerable to injury or loss. It is fortunate, however, that whipped antennas are capable of being regenerated; but in the meantime a lobster so affected will have to carry on without the benefit of these important sensory organs.

The whipped antenna, as shall be seen, is also brought into play during the pre-copulation ritual involving a soft-shelled female lobster that has just recently molted and a hard-shelled male lobster that she would like to pair off and mate with.

The Mouth Parts

There are several joints, limbs, and appendages that make up the mouthparts of the lobster. They are all very complex structures and perform various functions necessary for the processing of food and for movement of that food for consumption by the lobster. Most of the lobster's mouthparts are abundant with delicate sensory hairs, which serve as chemical and touch receptors that are capable of detecting the presence of food in the lobster's immediate environment. But the main task to be performed by the mouthparts is that of processing food. This is a very interesting process to watch in a controlled environment: once the lobster's sensory organs detect the presence of food, the dactylus of each of the large claws spreads wide open, the various mouth parts go into a frenzied activity, and the first pair of walking legs become increasingly active. The lobster at this point is greatly excited and stimulated by the taste and odor of the food. It will very quickly locate the food, which will then be drawn into the mouthparts. All of the mouthparts will work on the food in a steady and persistent manner to crush, mince, and grind the food down to smaller particles so that the food can be ingested into the tiny opening of the mouth. The finely ground-up food particles will then pass into and through a short esophagus that extends directly into the stomach where even more food processing will take place.

The Walking Legs

Although the lobster possesses five pairs of legs, the 2nd, 3rd, 4th, and 5th pairs are the only true *walking legs*. As we have seen, the first pair of legs of the lobster are the large crusher and ripper claws, biologically referred to as the *great chelipeds,* which actually do not serve any ambulatory functions at all. There is an observable difference in the walking legs, namely that the first two pairs of legs terminate in weak double claws, while the last two pairs of legs terminate as simple dactylus. It might be further observed that the dactylus, propodus, and carpus joints of these legs contain sensory hairs. These parts may have as many as one hundred bundles of "brushes," with each bundle containing from fifty to one hundred sensory hairs.

The Abdomen, Pleopods, and Tailfan

A very hard chitinous shell encases the top and sides of the lobster's abdomen. The underside of the abdomen features a thin membrane to which are attached the *pleopods, somites,* and *somite bars*. In addition to the fleshy and muscular mass of the abdomen, there is an intestinal vein that travels its entire length and terminated on the underside of the *telson*.

The pleopods, commonly referred to as *swimmerets*, play an extremely important role throughout the life of the lobster. Five pairs of these swimmerets are positioned throughout the underside of the tail section (abdomen). These abdominal appendages resemble paddle-like blades that contain a profusion of sensory hairs, more in the female than in the male lobster. Unlike the female, the first pair of swimmerets in the male lobster have been modified and are void of any of these sensory hairs. Rather, they are hard, bony, and pointed, and are referred to as the *stylets*. In the scientific community, they are often called the *gonopods*.. In any event, they have been modified in the male lobster to facilitate the flow of sperm to the female's seminal receptacle during *copulation* (mating).

In the female lobster the swimmerets are especially important to the processes of reproduction. Prior to egg laying, the swimmerets continually beat back and forth to cleanse the underside of the abdomen that will receive the extruded eggs. Following the fixation of the egg mass, the lobster will periodically energize the swimmerets in order to rid them of bottom sediments and marine parasites. This periodic cleansing also provided a continuous supply of oxygen for the growing egg mass. Much more will be said of these processes in Chapter 3.

The *tailfan* of the lobster is buttoned onto the sixth abdominal segment and consists of five relatively flat blade-like appendages. The posterior end of the lobster's intestinal vein terminates in the *anus*, which is located on the middle segment of the tailfan and on the bottom side of the *telson*. It is important to note that most of the outer edges of the tailfan are greatly imbedded with very fine sensory hairs.

The tailfan serves several functions that are of importance to the lobster. In concert with the tail section itself, the tailfan provides backward mobility of astounding speed; it is used to cover the pocket of eggs in a *berried* female lobster, and serves as a point of excretion for solid waste material. It is also common knowledge that the strong tailfan is used extensively as a "shovel" when the lobster is making a burrow in an ocean bottom consisting of either sand or mud.

The Brain And The Heart

The brain of the lobster consists of a small, almost flat white mass about the size of a dime, which is located directly under the shell of the carapace and directly to the rear of the rostrum. The heart of the lobster is a one-chamber organ about the size of a quarter. It is whitish in color, is also located under the shell of the carapace, and centered directly beneath the *longitudinal furrow line* of the lobster's carapace.

The Stomach

The stomach of the lobster is made up of a three-chamber organ, one chamber of which operates in a manner similar to a "grist mill." The stomach of a 1½ pound lobster would approximate the size of a silver dollar. It bears a whitish appearance and is capable of holding a considerable amount of food. The stomach is also located under the shell of the carapace and is surrounded by the many folds that comprise the lobster's liver.

The anterior chamber is used chiefly for the storage of food, while the posterior chamber has the job of sorting and straining that food. In between these two chambers is the amazing *gastric mill*. In the scientific community, this is referred to as the *gastroliths*, a fairly large bony-like structure that is brown and V-shaped. On either side of this bony plate is a row of lateral teeth that come up against the plate to work on the food particles that are now in the middle chamber of the stomach.

The stomach of the lobster is capable of quickly masticating food for use throughout the body as well as for eliminating solid waste material via the alimentary tract.

The Liver

The lobster's liver is undoubtedly the largest of the animal's organs. Positioned under the shell of the carapace, it consists of greenish folds that entwine themselves around and about the other organs located in this section of the lobster. The liver of the lobster is commonly referred to as the *tomalley* and is greatly favored by many people who are fond of the eating of the lobster.

The Gill Chambers

On each side of the lobster's section, and beneath the shell of the carapace, lay the *gill chambers*. Therein lay twenty pairs of gills that are essentially the animal's circulatory and respiratory systems. The gill chambers have two main openings, one for the salt water to enter from the posterior end, and one for the salt water to exit from the anterior end. A continuous flow of salt water passes over these gills and washes them with a fresh supply of oxygen, which is absorbed into the bloodstream. The deoxygenated water is then expelled from the gill chambers by the beating action of the *gill bailers*.

The Green Glands

Located at the bases of the 2nd antenna are the so-called *green glands* of the lobster. These glands are part of the lobster's excretory system and are used by the animal to rid itself of nitrogenous waste material. When signaled, the glands open up and any waste material is squirted out through the tiny holes at the bases of the 2nd antenna. Some lobster marine scientists suggest that this body function is more noticeable when a lobster is about to molt. They claim, in fact, that a lobster that is just about ready to cast off its old shell will first clear its system of this fluid waste material and that the fluid is often "squirted" a considerable distance.

The Ovaries

The ovaries of the lobster are composed of two cylindrical-shaped rods, or folds, that when fully extended can transverse about two- thirds of the lobster's length. This means that the ovaries can extend from the anterior end of the stomach to as far back as the 3rd, 4th, and perhaps the 5th abdominal segment at the time of maturity. During the period of development, the ovaries will present various changes in color: from bright yellow, to flesh salmon, to bright olive-green. After the female's first egg laying, the ovaries will take on a pea-green color, which will become progressively darker with age. Most people refer to the ovaries as the *coral*, and along with the *tomalley*, consider it as a delicacy.

Robert Delano Martin

The Sexual Differences

To the casual observer, two lobsters held side by side will look very similar to each other. But one of those lobsters "could be" a male and the other lobster "could be" a female. Although one might not really care about the sex of a lobster, the determination as to whether a lobster is a male or female can be made in several ways. An interest in the sex of a lobster will, however, afford the observer the opportunity to examine even more closely its anatomy

One way of determining the sex of a lobster is by examination of the *sternal somite bars* that extend horizontally across and along the underside of the lobster's abdomen (tail section). The female lobster will show, at each of the bar's center-points, a spur that is short, straight, and blunt. In the male lobster, however, each spur will be longer, slightly curved, and sharp. Another indicator of sexual difference is that the female lobster will feature a *bright blue shield* at the *oviduct openings,* which is the female lobster's seminal receptacle.

But these are far from being the only two ways in determining a lobster's sex. There are other ways, some subtle, and others immediately obvious. The subtle difference is that, in the female lobster, the lobster's top shell covering the abdomen (tail section) will have the appearance of being wider than that of a male lobster; the shell will also be more down-sloping along its sides than that of a male lobster.

Without a doubt, probably the easiest way to determine the sex of a lobster is by flipping it over and making a cursory inspection of the *swimmerets.* If all of the swimmeret's sensory hairs are soft and feathery, then that lobster is a female. To the contrary, if the first pair of swimmerets is of a bony-like structure and void of sensory hairs, then that lobster is a male.

For those who might be fortunate to observe the results of a typical trap-haul aboard a lobster boat, there are two occurrences that might be watched for that would definitely denote the sex of a lobster as being that of a <u>female</u>. The first is that the lobster fisherman will usually turn the lobster over to look for the signs of the presence of eggs. The second would be the lobster fishermen's quick inspection to see if one of the appendages of the *tailfan* bears a *V-Notch* which tells him that at some time in the past that particular lobster bore eggs, was V-Notched, and returned to the ocean as a conservation measure. Much more will be said about *berried* lobsters and V-Notching in the chapters that follow.

The material in this chapter has intentionally been kept brief and void of extraneous and unnecessary detail in order that the reader might gain a quick and easy comprehension of the anatomy and biology of the American lobster. The information contained herein, however, is considered paramount to the understanding of the animal's habits, life cycle, and reproduction processes, and so forth, that are covered in the pages to follow, especially so in Chapter 3, "The Lobster's Processes of Reproduction."

Chapter 3. The Lobster's Processes of Reproduction

The lobster, like other crustaceans such as crabs, shrimp, and crayfish will from time to time undergo a *molt,* or a shedding of the shell. The molting process marks both an incident and expression of growth which signals the biological fact that the lobster has outgrown its inelastic shell which now must be cast off. Stated in another manner, the interior fleshy body mass has grown while the shell that encases it has not.

A lobster may molt as many as ten times during its first year of life, including about five molts during the 20 to 30 day larval period. The lobster will molt periodically, but less frequently as it becomes of sizable length and weight that would earn it the right of being classified as a *jumbo lobster.* A jumbo size lobster (generally over 3½ pounds) examined in a lobster pool or fish market will often show evidence of its longevity and that it has not molted for a long period of time; it will be viewed as a lobster with barnacles and other marine organisms affixed to the large claws, the carapace, and other parts of the animal's shell structure. The longer inter-molt period of such a large lobster has allowed sufficient time to pass for an abundance of these organisms to become firmly attached to its shell and even to some of the appendages such the antennae, the walking legs, and the various mouth parts.

During a period of time prior to a molt, the lobster will prepare itself for the fete by absorbing lime and other minerals into the *longitudinal median furrow line* that extends from the beak of the rostrum to the edge of the carapace at the point that it meets with the shall of the tail section (abdomen). In a hard-shelled lobster, this line is as straight as an arrow! The other main point for the absorption of minerals is the *cervical groove*, which is an irregular depression that can be easily observed on the shell of the carapace.

During the cold winter months preceding a molt, any broken limbs or other injuries will be repaired. The lobster will be less active and more sluggish, and its total supply of energy will be devoted to the repair and regeneration of lost body parts. While lobsters molt in far greater numbers during the month of June through September in New England waters, the molting process is not strictly confined to these months and may, indeed, occur during any month However, lobsters going through a molt other than during the four months specified will be few in number as compared with the rest of the molting population.

The molting process is a very crucial event for the lobster. And sometimes it can be a fatal experience! The molting process can also result in the loss of limbs, such as the large claws, and can lead to deformities in the newly acquired shell.

The casting off of the *exoskeleton* occurs when the seawater has warmed and when other biological and environmental conditions are just right for the lobster. These conditions include an increase in the lobster's activity, an increase in appetite, the availability of food, and other factors that are favorable to the development of the new shell structure after the molt has taken place and the old shell has been cast off.

In preparation for a molt, the new shell will gradually be formed under the old shell. The molting process can best be described as having four distinct stages:

    1. The building-up of the new shell under the old shell to be cast off,

2. The actual shedding of the old shell,

3. A sudden increase in size,

4. A gradual hardening of the new shell.

The molt of a small lobster usually takes only a short period of time, perhaps from five to thirty minutes, and will occasion a truly unique experience in that the entire shell structure will be cast off as one piece. It is an accepted fact that the larger the lobster, the longer will be the period of time for the lobster to undergo and complete a molt. As compared with a smaller lobster, it takes a jumbo lobster a long time in order to wriggle out of its heavy, tough, and old shell that has served as a coat of armor for that lobster for a good number of years.

A lobster's disposition will change just prior to a molt. At this time, there will be an attitude of activity, excitement, and uneasiness. Francis H. Herrick, quoting an account by Anderton, describes this pre-molt activity of one lobster under study:

> "A male lobster was seen to be behaving in a very peculiar manner in the shallow end of the pool. It would walk alongside the concrete dividing wall for a distance of about 5 feet, halt, and then turning around would retrace its steps the very same distance in the opposite direction. In this manner a rut several inches deep was formed in the gravel and at one end of this the lobster scooped out a large hole about 4 inches deep and 12 inches in diameter. The water had to be temporarily withdrawn from the pond, but as soon as permitted to do so this lobster resumed its peculiar walk, and continued through the night and the following day."

In its natural habitat in the wild, the lobster will usually cast off its old shell during the twilight or early evening hours. It will depart from its old protective covering by withdrawing the anterior part of the body backward, and by withdrawing the posterior part of the body forward; and all of this being accomplished in such a manner that both sections exit through a rent in the soft membrane between the shell of the carapace and the shell of the tail section. And, according to Herrick, the lining of the stomach and esophagus come out by way of the mouth, while whatever else is discarded through the alimentary tract.

We have seen that a lobster soon to go into a molt will exhibit several telltale signs that indicate that the event will soon take place. But once the molting process begins, *there is virtually nothing that can be done to either stop or reverse the process!* Herrick has described the molt of a lobster in the following meticulous detail:

> "When the lobster is approaching the critical point, the carapace, or shell of the back, gapes away a quarter of an inch or more from the tail. Through the wide chink thus formed, the flesh can be seen through the old and new cuticle, giving it a decidedly pinkish tinge. Take the lobster up in the hand now and the tail drops down as in death, the strong muscles which bind the pleon to the carapace being completely relaxed. When this stage is reached the time of exuviation is at hand and the process becomes purely automatic, the animal having no control over its own movements.
> "The period of uneasiness, which foreshadowed the molt, and was very marked, ended in this lobster by its rolling over on its side, briskly moving its legs, and bending its body in the shape of a V, the angle of the V corresponding to the gaping chink between the dorsal shield and tail. Presently the old cuticle, holding these parts together, began to stretch, the wall of the body pressing against it with considerable force, and the hinder end of the shield being slowly lifted up, while the anterior part remained attached to the rest of the skeleton. The slow but sure pressure of the parts within cause an increasing tension in the yielding cuticular membrane, which finally bursts, revealing the brilliant colors of the new shell. The legs and other appendages are occasionally moved, but no marked convulsive movements are to be seen. The carapace has

now become raised to an elevation of perhaps 2 inches in its hinder part, in consequence of which, the anterior end being fixed, the rostrum is bent forward, and the animal presents a very singular appearance.

"When this stage has been reached the lobster becomes quiet for a few seconds and then resumes its task with renewed vigor. From this time on until free, its muscles work intermittently. The doubled-up fore part of the body, with each effort by the animal, is more and more withdrawn from the old shell, and this implies the separation of the skin from the intricate linkwork of the internal skeleton, and particularly in its release, together with part of the nerve cord, from the closed archway of structure, as well as freeing of the 28 separate appendages from the old cases and tendons, for the accomplishments of which several adjustments are made in advance. The cuticular sheath of every ectodermic structure is ripped off. The exoskeleton formed to fit so complicated a mold in virtually a continuous structure, and from the method of its regeneration, the sloughing of one part necessitates the shedding of the whole."

Herrick concludes with his final observations:

"The carapace is now elevated to such an extent from behind that the rostrum is directed downward and backward. The lobster is still lying comparatively quiet upon its side, but the muscles of its appendages are undergoing violent contractions as the animal tugs and wrestles violently as if to free itself from ropes which bind it down firmly on every side. The carapace is unbroken, yet the two halves bend as upon a hinge along the median line. Presently the pressed-down bases of the antennae, the eyestalks, and the bent-down rostrum of the new shell can be clearly seen. No part of the covering of the large claws or any of the legs have been split or cracked. The muscular masses of the powerful claws have been withdrawn through their narrow openings without a rent.

"Finally, a few kicks free the entire forward half of the body, the antenna, chelipeds, and the various other parts, which now lie above or to one side of the old covering. The tail has been gradually been breaking away from its old case, and as soon as the forward part of the body is withdrawn the lobster gives one or two final switches and is free."

The part of the molting process that seems to puzzle most anyone interested in the lobster is that of the meaty flesh of the large claws being capable of being withdrawn from within the confines of the old shell, and without any breaking or splitting whatsoever. It is a truly remarkable fete inasmuch as the claw at it widest point is roughly two to three times larger than the widest point of the merus and carpus appendages (knuckles). The answer to this puzzling question apparently is contributed to the elasticity of the muscles of the large chelipeds, the contraction of the muscle mass, and the ability of the lobster to withdraw fluids from this muscle mass. Once the muscle mass has been contracted, and the fluids have been displaced, the lobster is able to wrestle the shrunken mass through the carpus and merus appendages, and doing so without any cracking of the shell of the large chelipeds at all.

One of the most interesting aspects of the molting of the lobster is that the shell that has been cast off is an exact duplicate of the newly formed shell. In fact, in the moments following a complete molt, one would find it very difficult - without looking very, very closely – to identify which of the two specimens is that of the old shell and which is actually the newly molted lobster. Placed side-by-side, they appear identical!

Following a molt, one of the very first meals for that lobster will be the old shell that has just been cast off. The newly molted lobster will often pull the shell into a burrow or into some hole of seclusion to feast on it. The lime (calcium carbonate) and other minerals contained in the shell will be an important attribute for the prompt development of the lobster's newly acquired shell structure. The new covering of chitin is found to be quite slippery, and if the lobster is taken in hand, it will feel limp and rubbery-feeling.

The shell of the newly molted lobster will require about six to eight weeks before it attains the hardness of the old shell. The lobster, during this period of time, will not have the strength or agility to pursue rigorous activity, but will be capable of moving about by flipping its strong and muscular abdomen.

[It is most likely apparent to the reader at this point, and even at the beginning of this chapter, that the material presented in Chapter 2 was placed there for the purpose of making the material now being presented more easily comprehendible. In this regard, perhaps a reading – or re-reading – of that chapter concerning the anatomy and biology of the American lobster would be both appropriate and beneficial for an understanding of the contents of this chapter which addresses the important processes of reproduction: the molting process, the mating process, and the egg-laying process – Author]

The Mating Process

The mating and copulation of the male and female lobster will only take place after sexual maturity has been attained *and only when* the female lobster is in a soft-shell condition immediately following a molt.

Sexual maturity in the female lobster rests largely with the degree of development of the ovaries and the *ova* (the eggs). Lobster marine scientists generally agree that there are three basic stages of ovarian development: immature, developing, and mature. Ovaries that are white or creamy in color and with ova that measure 0.4mm or less in diameter are classified as being immature; ovaries that are yellow or orange and with ova measuring between 0.5mm and 0.7mm in diameter are classified as being in the developing stage; and ovaries that are dark green and with ova of 0.8mm or greater are considered as being mature.

The results of extensive scientific research indicate that most female lobsters do not begin to become sexually mature until a *carapace length* of about 3-3/16 inches (about 80mm) is attained. As a conservation measure, and in order to allow for the female lobster to reproduce at least once before being removed from the fishery, the lobster gauge in all lobster-producing states has been increased to 3¼ inches. Lobster marine scientists generally agree, however, that virtually all female lobsters with a carapace length of 3-7/8 inches can be considered to be sexually mature.

Male lobsters mature earlier than do their female counterparts. The main criteria to establish sexual maturity in the male lobster is the presence of sperm in the *vas deferens* and *testes* which are located at the bases of the last pair of walking legs. Studies by research personnel, and especially by Jay S. Krouse of the Maine DMR, indicate that approximately one-half of the male lobsters studied had sperm at about 44mm carapace length measure, and that the males above the 55mm carapace length measure were almost all with sperm.

Being sexually mature, and having just recently molted, the female lobster will be strong enough to seek out a mate to "pair off" with. It is a widely held opinion that the female lobster will prefer to mate with, and achieve greater success, with a male lobster that is larger in size.

So, the stage is set for the female lobster, shortly after molting and in a soft-shell condition, to immediately commence a search for a larger male lobster to mate with. In the process of doing so, she will excrete a substance, known as a *pheremone*, to attract and entice any male lobsters in the immediate ocean-bottom environment, one of which she will choose as her mating partner.

In his classic work, the *Natural History of the American Lobster,* Francis H. Herrick describes a *pre-copulation ritual* that was observed and reported by Anderton. One particular female lobster was observed laying for a period of time beside her cast-off shell, and then:

> "Two hours afterwards it was seen roaming around the pond and frequently approaching the various shelters, returning regularly and fearlessly to a shelter containing a large male. On approaching the entrance to this shelter the large claws were extended in a direct line with the body and the antennae were thrust within the shelter. After a few moments the rostrum of the male appeared, the female meanwhile rapidly whipping her antennae across the now projecting rostrum of the male, which in turn showed increasingly signs of excitement, the antennae being whipped very rapidly over the female in the same manner.
> "After an interval of perhaps a minute the male gradually withdrew from his shelter, the female at the same time turning on its back. *Coition* took place at once, the act occupying only a few seconds, the male retiring at once to its own shelter and the female into another one. The following day both were observed living in one shelter, and they continued to do so, off and on, for several weeks."

Anderton, responding to specific questions put to him by Herrick, further described the mating process he observed:

> "The female lobster after casting [does] appear to seek out a male lobster as soon as the distressing effects of molting have somewhat worn off. Male and female have frequently been observed living in one shelter for some days and even weeks after coition. The act of coition is very brief, and will not occupy more than half to a full minute. They copulate, as we express it, "belly to belly," and head to head. The large chalae do not come into use

during the act as far as I have observed. The female voluntarily turns over almost completely onto her back, the excited male completing the process for her."

The Egg-Laying Process

The natural progression of events in the lobster's reproductive process follows a very consistent and predictable pattern. Having delved into the aspects of molting and mating, and the act of copulation itself, attention must now be focused on the female lobster, the laying of her eggs, and the fertilization of those eggs by the sperm of the male lobster, being deposited within her during some previous mating experience.

The vast majority of female lobsters who are going to lay (spawn) eggs in any given year will perform this fete during the summer months. It is estimated that 80% or more of the spawning activity will take place during these summer months, with the remainder occurring during the fall and winter months. A two-year period elapses between each egg laying of the female lobster. Stated in another way, the typical female lobster spawns on a two-year cycle, or every other year.

Prior to laying her egg mass, the caring mother will be very meticulous in the performance of several hygienic chores in preparation for the egg-laying process that will soon take place. She has been seen to prop herself high up in a tripod-like manner by supporting herself with the large claws at one end and by the tailfan at the other end. In such a position, and with the body slightly elevated, she will scrape, pick, and brush away any foreign sediments and parasites that have accumulated on the swimmerets.

As was the case for a lobster undergoing a molt, the egg-laying process in the female lobster will most likely occur during the evening or early morning hours. In her solitude, she will commence the event by rolling over onto her back; she will then arch her body section up as high as possible, while at the same time forcing her large claws, like a pair of crutches, into the bottom substrate With the posterior part of the tail section arched and curled back towards the body, the female is now properly positioned to extrude the eggs, not as a solid mass, but only one at a time.

The *oviduct* openings in the female lobster are positioned between the second and third pair of walking legs, one tiny opening being located on each side of the hard bone of the *sterna*. It is through the oviduct openings that the eggs must pass through during *oviposition*. Before the eggs are extruded from the oviducts, or at some moments shortly thereafter, they become covered with a cement-like adhesive substance that enables them to become affixed to the swimmerets and to one another in a solid mass. These are several theories regarding the source of this adhesive substance. One theory is that the substance originates in the linings of the oviducts and attaches to the eggs as they are extruded; another theory is that the adhesive substance originates in the *tugmental glands* of the lobster's swimmerets; and still another concept suggests that the substance originates from the linings of the abdomen. Regardless of the origin, however, without the presence of this adhesive substance, the extruded eggs would simply "float away" and insemination would not take place.

Francis H. Herrick describes this egg-laying process in quite some detail:

> "The lobster turns over onto its back and by the aid of the two large claws and ridge of the abdomen makes a tripod of herself,

the head being considerably more higher than the posterior portion. The abdomen is then strongly flexed, forming a pocket, and the satae on the edge of the abdominal segments make the space along the sides perfectly tight. An A-shaped opening into the pocket is formed by the telson and the sixth abdominal segment. This opening, when the abdomen is flexed, is slightly posterior to the first pair of swimmerets. The eggs then flow from the two genital openings in a continuous stream, one at a time, and pass along at the bases of the last pair of walking legs and into the "pocket." The course of the eggs into the "pocket" is further assisted by the constant pulsing of the first pair of swimmerets, causing an in draft, which carries them rapidly inside.

"None of the eggs are lost on the passage from the genital openings to the "pocket" unless the lobster is disturbed. As the eggs leave the oviduct they become covered with an adhesive substance, which causes them to stick together and to the swimmerets.

"The period of oviposition in the lobster under observation was just over four hours. Half an hour after the eggs had ceased to flow the lobster righted itself and walked into a corner of the tank, eventually getting into a nearly perpendicular position, with the head downward. It remained in this position for the rest of the day. Next day it was walking around the bottom of the tank in the usual way of a berried lobster."

Following the extrusion and fixation of the eggs to the swimmerets, the egg-bearing lobster will be a good mother to her brood for a period of 10 to 11 months during which time the embryo development takes place. During this long fosterage period, the berried female will guard her brood "like a hawk" by instinctively folding her abdomen and tailfan in, under, and towards her body section to form what Herrick has called the "pocket." She must be especially wary of any and all of her natural enemies in her surroundings, such as the codfish (probably enemy #1), sculpins, sand sharks (dogfish), and other forms of marine life that are ever-present and extremely adept at scooping off the eggs of an unwary berried lobster. Periodically, however, the female will unfold, lift up, and straighten the tail section.

While in this position, she will rapidly move her swimmerets back and forth to perform a cleansing function, and thereby shaking loose any parasitic organisms and bottom sediment that have become attached to the ova and the swimmerets. The lobster will also, in this position, be aerating the ova by providing it with a fresh supply of oxygen. But once these "housekeeping chores" have been carried out, the tail section will once again be immediately tucked under to protect the pocket of eggs.

Freshly extruded eggs are dark brown in color, and when viewed as a mass may take on more of a brownish-black color. As the ova approach the end of its development, examination of any one of the thousands of eggs will reveal two iridescent spots. These spots are the lobster's eyes.

Research over the years indicates that the number of eggs laid will be somewhat proportional to the volume of the female's body. Therefore the larger and older female lobster will have the *fecundity* (egg-bearing capacity) to lay and foster a greater number of eggs than would be the case for a smaller female lobster. A female lobster at the time of her first egg laying might produce something in the order of 5,000 eggs. On the other hand, a female lobster that is much older and larger may extrude as many as 90,000 eggs or more.

The Hatching Of The Mature Eggs

Upon full development and featuring a golden-brown color, the eggs are ready to be hatched into the *water column*. It has been observed that just prior to the first release of eggs, the lobster is marked by a state of restlessness, an expression of uneasiness, which no doubt is brought upon by the subtle movement of the thousands of eggs affixed to the underside of the abdomen. The female lobster apparently feels this movement, this stirring, which signals her that the time is close at hand for release of those eggs into the water column. She will then lift her body high up off the bottom and will maintain her tail section in a highly elevated position. With her body so stabilized, she will vigorously shake her entire body, and with the rapid movements of the swimmerets, her brood will be dispersed into the water column. One direct observation, as reported by Francis H. Herrick, helps us to understand this very important part of the female's reproductive process:

> "Toward seven or eight o'clock in the evening, the female commenced to stir herself in her prison by presenting an attitude altogether unusual and characteristic. Her feet are stretched out almost rigid, her tail extended to the full in a horizontal direction, forming, with the rest of her body a nearly straight line. She walks, we might say, upon her toes, so careful is she to hold her entire body as far away as possible from the bottom of the aquarium. This feat lasts for a certain time; then quickly lowering her head and the fore part of her body until she rests upon the ground with her outstretched claws, with the tail at the other end raised at an angle of 45 degrees and kept stretched, we see her violently shake her swimmerets with such rapidity that the eye cannot follow the movement, and a veritable cloud of larvae are sent far to the rear and dispersed in all directions.
> "This phenomenon lasts from 15 to 20 seconds, and the female thereafter returns to her habitual attitude, to depart from there no more until the following evening."

The female lobster does not necessarily release and hatch out her brood during just one single event. The hatching of the larvae might occur over as many nights as required to free all of the larvae. This, no doubt, results in a far greater number of the larvae to survive in the water column on their way to the surface where they will commence their *free-swimming* pelagic life.

Life In The Water Column

A larval lobster upon hatching is a very small creature – a mere 1/3 of an inch in length! But it shall, as it must, face the perils of the ocean and a totally new, strange, isolated, and vulnerable environment. After finally being shaken free from the cuticle that imprisoned it, the tiny lobster emerges as a *free-swimming* creature that enters the *water column* and gradually rises to the surface. It is there that the larval lobster, if it is fortunate enough, will endure a pelagic life and pass through metamorphoses consisting of five molts.

The first molt of the larval lobster actually occurs just prior to hatching. It is the release of the larval cuticle that releases it from the shell membrane that has encased it for several months. Therefore, hatching, molting, and the unsheathing of the swimmerets are coincident and must take place if the minuscule larval lobster is to successfully free itself of its encasement and commence to rise in its upward voyage to the surface of the ocean.

Larval lobsters on or near the surface of the ocean water are common prey to both fish and birds, and because of their extremely cannibalistic nature at this time of their lives, any weak or dying among their lot will be quickly set upon and devoured. Any larval lobsters that are unhealthy, lack vigor, and are drifting aimlessly on or near the surface will be immediately seized and eaten by their numbers. Larval lobsters drifting here and there are also vulnerable to the whims of the ocean, and are therefore capable of being easily washed ashore or thrown up and stranded on outcroppings of ledge that are commonplace along the North Atlantic seaboard of the United States and Canada.

The diet of the larval lobster consists of essentially any organism, dead or alive, that is in suspension on or near the surface of the water. They exhibit a marked tendency to attack and seize any object. It is for this reason that in a controlled environment, such as at a lobster hatchery, that they are suspended in cylindrical-shaped *rearing tanks* where the seawater is kept constantly swirling and in motion.

The larval lobster – if it should somehow manage long enough to reach the bottom-seeking stage – will undergo a series of metamorphoses, or stages, each of which will be preceded by a molt. Thus, the molting upon release from the cuticle at birth will introduce the *second stage* lobster, the second molt will introduce the *third stage* lobster, and so forth. During the typical 15 to 30 day larval period, the lobster will pass from one stage to another, molting each time, and will show subtle changes in size and appearance. Upon reaching the fourth molt, the larval lobster will take on a sudden leap in development. Unlike the previous stages, the fourth stage lobster appears to become a new animal and, for the first time, resembles the shape and form of the lobster. Herrick makes these observations about the fourth stage lobster:

> "In form, color, habits, and instincts it (the fourth stage lobster) differs strikingly from every preceding stage... it swims on the surface with greater agility, precision, and speed than at any former stage... the lobsterling glides swiftly along by the action of its swimmerets... the big claws are extended straight out in front of the head and held close together... the great chelipeds are

> long, slender, and end in symmetrical claws of the toothed type... the exoskeleton is now re-enforced for the first time with considerable deposits of mineral salts, especially lime... it is quite translucent... and the body of the lobster is studded with sensory hairs..."

According to Herrick, "the preying instinct is more marked, and the fighting instinct, the instincts of fear, "feigning," and hiding are all developed at the beginning of this stage or in the fifth, which follows, when the animal goes to the bottom to stay."

The fourth stage lobster will have grown in size from about 1/3 of an inch to about ¾ of an inch. Its color will have changed from a translucent white at birth to a beautiful red; it will have developed its body structure and body functions to the extent that it can now take on all of the characteristics, including behavior, in somewhat the same manner as an adult lobster.

The extremely low survival rate of larval lobster in the wild has always been of a major concern to lobster marine scientists. In fact, very few of these larval lobsters hatched by the female on the bottom of the ocean will ever survive to the critical fourth or bottom seeking stage. It is the estimate of lobster marine scientists that _**less than one-tenth of one percent (< 0.1%)**_ will ever reach the fourth or fifth stage when they will go to the bottom and live. This extremely high mortality rate can be compared with the more favorable survival rate reported by marine scientists, such as at Martha's Vineyard, who estimate that the survival rate for their hatchery-raised lobsters is approximately 35% If both of these estimates are reasonably correct, and assuming that a lobster hatched out 6,000 eggs, then not more than 6 of those eggs hatched in the wild would survive to the bottom-seeking stage, whereas 2100 of those eggs hatched at a lobster hatchery would survive to the bottom-seeking stage! It would seem, therefore, that a long-term effort might be made at the hatchery to raise more and more larval lobsters to the bottom-seeking stage, and then release them at carefully selected locations off the coast where they will immediately descend to the bottom to stay and live.

Swimming stages of larval lobsters on or near the surface of the water. The cannibalistic attitude of the larval lobster is demonstrated in b. The final stage larval lobster is shown "going to the bottom" in f. From Francis H. Herrick, "Natural History of the Lobster."

The Tale of the Lobster

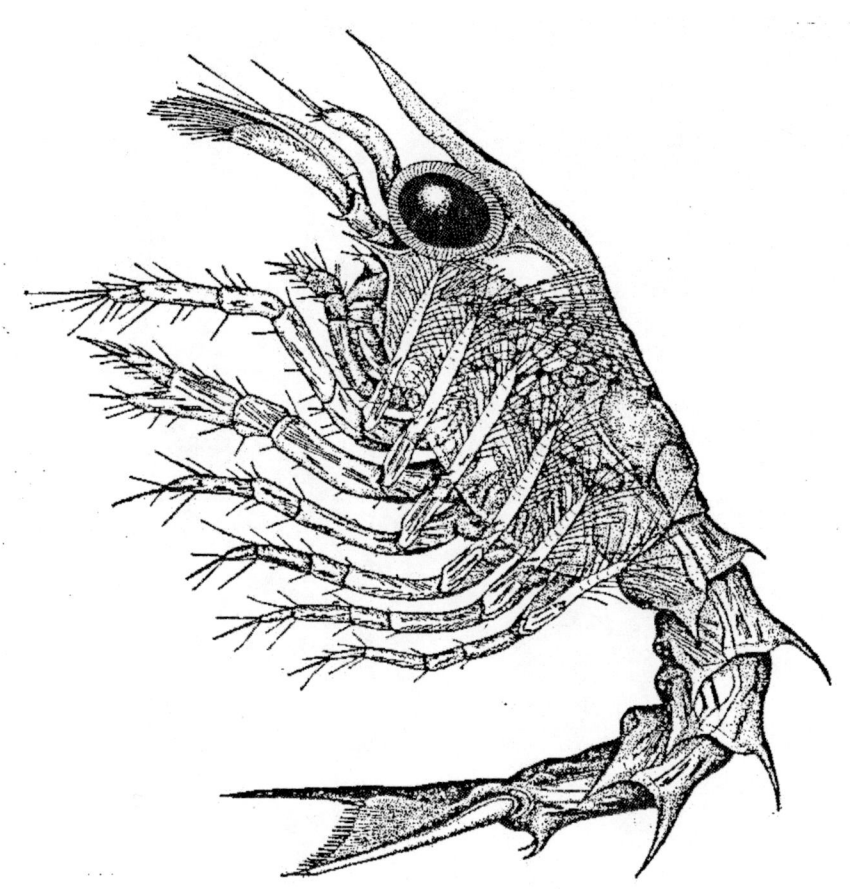

The first stage, free swimming stage, of the American lobster. Length: About 8mm, or a little less than 1/3 inch. Taken from Francis H. Herrick, "Natural History of the American Lobster."

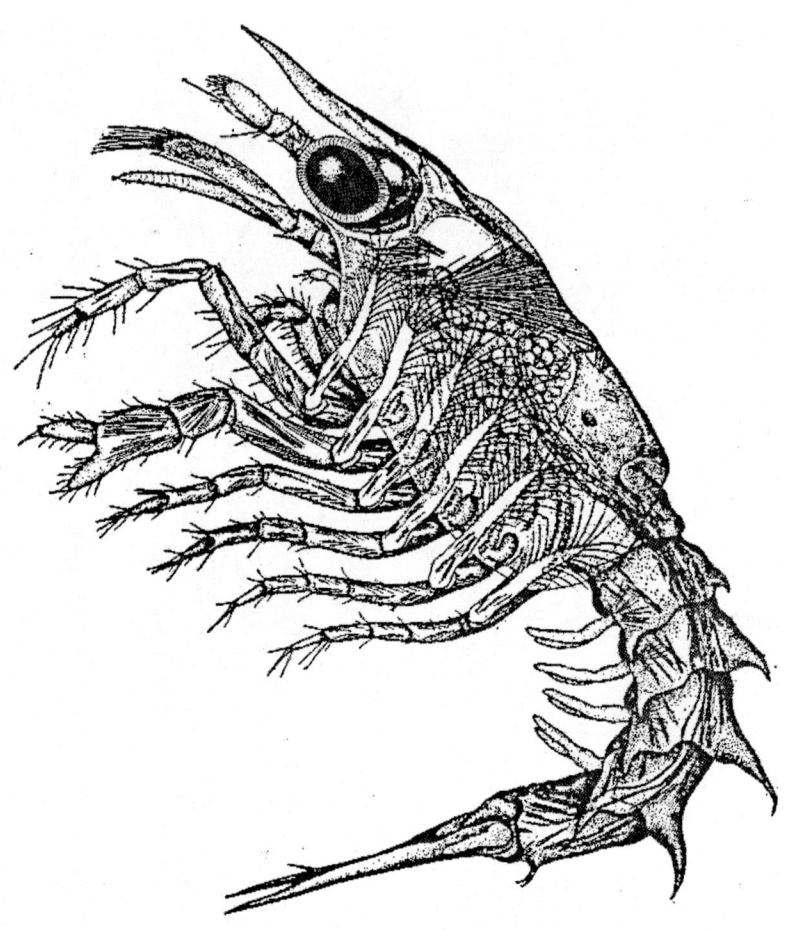

The second stage, free swimming stage, of the American lobster. Length, 9mm, or 0.35 inch. Taken from Francis H. Herrick, "Natural History of the American Lobster."

The Tale of the Lobster

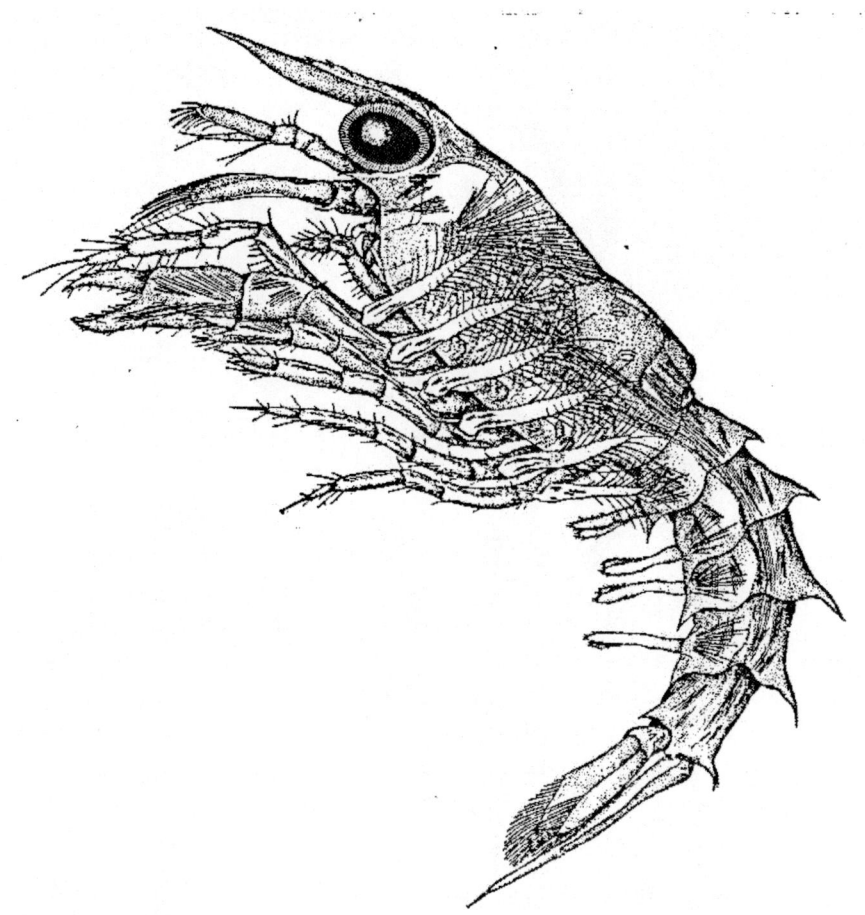

The third stage, free swimming stage, of the American lobster. Length, 11.1mm, or 0.44 inch. From Francis H. Herrick, "Natural History of the American Lobster."

Robert Delano Martin

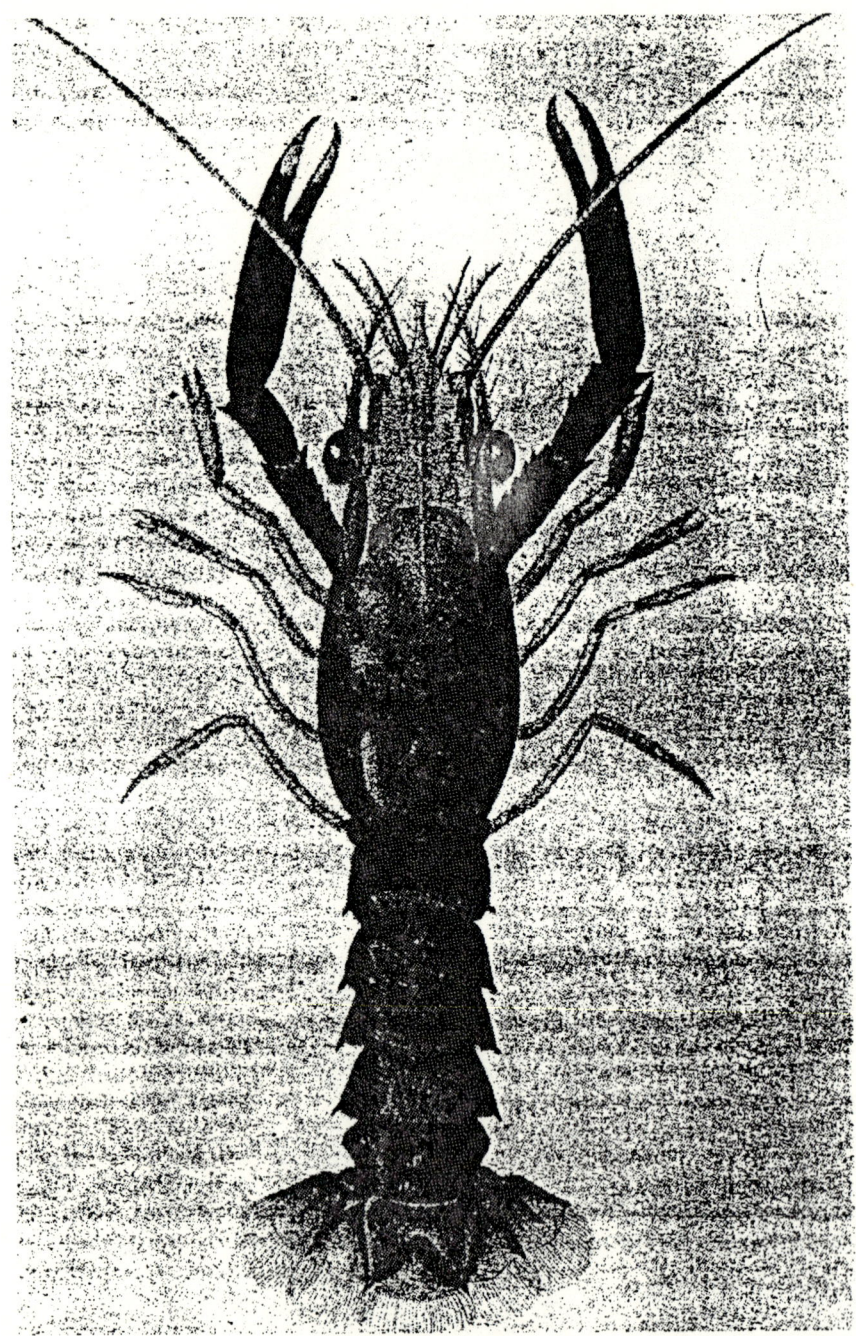

The fourth stage of the American lobster. Length, 14.6mm, or a little more than ½ inch. Taken from Francis H. Herrick, "Natural History of the American Lobster."

The sixth stage of the American lobster. Length, 16.0mm, or 5/8 inch. Taken from Francis H. Herrick, "Natural History of the American Lobster."

Chapter 4. Life On The Bottom

We have seen that the less than one-tenth of one percent (< 0.1%) of the larval lobsters that survive their pelagic life on the surface and make it to the bottom are obviously an extremely important part of the stock that contribute to the survival of the North American lobster fishery. These "survivors" will immediately seek out a place of hiding and seclusion upon reaching the bottom. Their new abode will be some sort of shelter that will be suitable for protection from their natural enemies that share the same space. The codfish, haddock,, and other types of groundfish such as sandsharks, eels, and sea ravens (sculpins) are mortal enemies of the lobster. *Anemones,* which are colorful plants anchored to the ocean floor, can also take a heavy toll on the small, unwary, and uninitiated *lobsterlings* who just happen to come into encounter with them.

The home of the lobsterling in its natural habitat will be in a rock crevice, under rocks, or in any artifact that is available to it on the bottom. When clay, mud , or gravel substrates exist, intricate tunnel systems become the hiding places and living quarters for the minuscule lobsterlings.

Not much is actually seen of the lobsterling during its first few years on the bottom. This is most likely attributed to its instinct of fear as a small marine specie, and will cause it to seek out virtually anything on the ocean floor in which to hide. The lobsterling might be found in a cut-off *ghost trap*, a tin can, a boot, or just about anything else that happens to litter the ocean floor. According to Dr. Richard Cooper, a marine scientist with the National Marine Fisheries Service at Woods Hole, Massachusetts, small lobsters go into immediate hiding and are difficult to locate for quite some time. At a Fishermen's Forum at Rockland, Maine, Dr. Cooper stated that "lobsters don't begin their nocturnal behavior until they are about 40 to 45mm carapace length measure. That's about half the size of a minimum size lobster. Lobsters up to about half that size of 3-3/16 inches* spend their entire life in the planktonic world on the ocean floor. They spend their daytime and nighttime in a very intricate tunnel system** on the ocean floor. And they won't come out!"

[Author's Notes: * 3-3/16 inches in carapace length measure. The minimum legal size for lobster is now 3¼ inches in carapace length measure.
** As previously discussed, there are artifacts on the bottom that could also serve as places of hiding for these lobsterlings.]

It is estimated that it takes approximately five to seven years for a lobster to attain the minimum legal size of 3¼ inches in carapace length, and assumptions have been made that a lobster will not commence its marauding tendencies until about the third or fourth year of life on the bottom. It will be at this time when the lobster will truly behave and act in the manner of the adult lobster. Its lifestyle will be given up to searching out food, eating, and hiding. It will prefer a rocky bottom with a gravel substrate and in an area where there is an ample supply of plankton, which, during the winter months, is considered to be the mainstay of the lobster's diet. If a rocky bottom is not available, but a muddy bottom is, the lobster is capable of "digging in" by burrowing into the mud and forming a tunnel system that often extends two to five feet or more. The animal will utilize its strong crusher and ripper claws and walking legs to burrow out the tunnel system. The powerful abdominal tail section and tailfan will be used like a "shovel" and "push broom" to clear the loosened-up substrate from within the tunnel and from the entrance to the tunnel.

During the daylight hours, lobsters are often found at the entrances to their shelters, with their long antennae outstretched and sweeping the water for the presence of food, pheremones, and even possibly an interloping enemy nearby. Upon the threat of a predator approaching, the lobster has the capability of reversing itself in a very expeditious manner. It will prefer the smallest shelter possible, but large enough to enable it to force the large claws into the walls of the shelter. In this position, and with the rostrum forced into the ceiling, it becomes virtually impossible to dislodge the animal and pull it out of its shelter.

When a lobster is caught out in the open, and is threatened with attack, its natural stance will be such that the large claws are extended above and slightly back, as in a spread-eagle fashion, and the long antennae are swept backwards in a direction parallel to the animal's body. The lobster will stand tall on its walking legs and with the tail section kept curled under and tucked under. The powerful abdominal tail section, sometimes casually called "the flipper'" is capable of propelling the lobster backwards in a very expeditious manner. In the event that one or both of the large claws are seized by a combatant, the lobster, as we have seen, has the unique ability to *throw* or *shoot* a claw in order to make escape possible. The lobster will give up a claw, or even both claws to escape a worse fate, up to and including being overcome by its attacker. To put it another way, the lobster will give up a limb in order to save its life! Many of the one-claw *cull lobsters* found in lobster pools, fish markets and supermarkets have gotten that way because of such dangerous encounters. However, cull lobsters could have lost one or both or their large claws for reasons unrelated to "throwing" them in combat. The careless handling of lobsters by lobstermen, their *helpers,* wholesalers, retailers, and others in the distribution chain all contribute to the large number of cull lobsters that are eventually sold to consumers.

The lobster is often referred to as an animal of the twilight. Continuous research over the years by lobster marine scientists using submersibles reveal that the lobster, like the landed raccoon and skunk, are far more active at night than during the daylight hours. The diet of the lobster is virtually anything, either dead or alive, that it is capable of being seized up and eaten. The animal is, after all, a scavenger. In addition to the bait in the lobsterman's traps, the diet of the lobster includes many species of fish, but it is very fond of crabs, clams, mussels, mollusks, and other types of shellfish. Probably the crab is favored the most of all. And as lobsters are great scavengers, it is a widely held opinion that they actually prefer fresh bait rather than old and decaying bait. Conventional wisdom is that the more putrid the bait, the better the catch, but this is probably not the case at all; rather, the lobster is merely attracted to the trap by the chemicals and odor of the putrid bait that are being dispersed in and around the trap. The lobster may approach the trap, but will probably not be enticed to enter, and will no doubt do a little more "night-walking" in search of a more scrumptious meal in a nearby trap that is more freshly baited. Since lobster research has shown that lobsters can subsist over long periods of time without virtually any food whatsoever, it should come as no surprise that they have a tendency to be selective and will find a trap that has been freshly baited, rather than a trap containing remnants of bait that is decaying and putrid. A casual observation of an active lobster on the bottom illustrates that it nimbly walks along the bottom with its tail section pretty much fully extended and straight out in a horizontal plane. The inactive lobster, however, will carry its tail section in a curled-up manner that is characteristic of the berried female lobster with eggs.

We have seen that the lobster is a very agile, wary, and pugnacious crustacean and that its life is preoccupied with either hiding or searching for food. In the wild, and when the lobster population is active, but when the food supply is inadequate, the lobster demonstrates its character of being cannibalistic. But when the saltwater temperature cools, the lobster will go off its food, become more sluggish, and seek a sedentary life that is spent in seclusion. If in a mud burrow, the lobster might be found with only the rostrum, stalked eyes, and whipped antennae protruding from the burrow. If in a rocky environment, the lobster might be found secluded in and under rocky crevices. However, the reversal of these tendencies will occur when the seawater warms and the lobster becomes more active, acquires a greater appetite, and becomes very aggressive in its search for food.

An Imaginary Voyage To The Bottom Of The Ocean

Let us now to take ***an imaginary voyage to the bottom of the ocean*** to observe the behavior of one particular lobster roaming about on the bottom of the ocean a half-mile off the rocky coast of Maine. It is nighttime and darkness is on the face of the deep. But that need not bother nor hinder us in any way, because we have been given *all-perceiving and omni-powerful eyes*, and our vision has the unique capability of permeating the gloom and darkness of the water.

Among the various forms of marine life, we spot a lobster, a beautiful 1½ pounder that is slowly walking along the bottom. It is walking nimbly on the tipity-tops of its slender walking legs, its tail section well up off the rocky and gravely substrate, and the crusher and ripper claws are extended straight out in the direction of travel.. The long branched antennae are in constant motion and are being steadily whipped back and forth through the water as if in search for something. It stops periodically, pauses, and then proceeds once again in a more deliberate manner. Then, all of a sudden, the lobster is becoming more excited, with the whipped antennae becoming more and more active, and with the various mouthparts moving and twitching in a very frenzied fashion. The lobster stops, so we shall stop, and as it proceeds again, we shall proceed with it. It meanders along cautiously and in the direction of one of the several traps that are resting on the bottom. It approaches one particular trap, and detecting the smell of the bait in the trap, decides to enter it.

As the lobster looks for a way to get into the trap, we are now in a position to make some observations. We study the lobster and come to the conclusion that it is a beautiful male and that it is perfect in every detail! Having decided that the sex is male, we think that instead of referring to the animal as "it," we shall give "it" a name. We shall call him "Homar."

Homar ever so slowly navigates his way around the trap while looking for a way to get inside. The water is dark and murky and the going is slow, but as he starts walking up the side of the trap he finds that it becomes easier going as he walks from one nylon strand to the next. Upon reaching a 6¼ inch round *ring*, he decides to crawl through it and into the trap. He then descends into the *"kitchen"* where the bait is stored on the bait spindle.

Having stripped apart and consuming as much of the *redfish bait* as he desires, Homar then moves about in the kitchen in search of a way out. He tries to exit the trap through the same hooped opening that he used to enter it, but because of the construction of the trap's *head* and *ring*, he cannot seem to figure out a way to crawl up and through the opening. Homar continues to walk around the inside of the kitchen in search of a way out. He then comes upon another opening on the other side of the kitchen, but escape is impossible because the nylon mesh is just as straight up and curved-up as the one he had just before unsuccessfully tried to exit through. Being very persistent, Homar again searches further and discovers yet another hooped opening in the meshed netting near the center of the trap. He crawls up and through the opening and plummets into the "parlor" section of the trap.

Homar probes around the trap parlor looking for a means of escape, but the only opening of any size is the one that he has just crawled through. He once again finds that escape is impossible because of the shape and the sloping angle of the meshed netting. He looks further and notices a small rectangular opening near the bottom of the parlor. Homar tries desperately to wriggle his body through the opening, but try as he may, he finds it impossible to get his large claws and body through such a small and narrow opening. But being of dull wit, he does not realize that this opening is an *escape vent* that was intentionally designed as an escape hatch for only smaller lobsters. Having exhausted all means of escape, Homar settles down and realizes that he has become trapped!

The Tale of the Lobster

It's getting brighter now and we notice a disturbance from up above. It must be a lobster fisherman who is beginning to haul his traps. And then, all of a sudden, we see the trap, with Homar inside, being jerked up off the bottom at a 45-degree angle. With a loud **WHOOSH,** the trap disappears from above us and is out of sight within a matter of a few seconds! We decide to go topside to investigate what's going on up there.

At Topside

Arriving at topside we are able to see but not be seen. We will be neither disruptive nor obtrusive to anyone or anything that is happening on the surface because we have been given the power to render ourselves *entirely invisible*. So we will just observe for a while.

We find that Lobsterman Sam has just hoisted the lobster trap up and onto the *gunwale* of his boat. He unhinges and swings open the door at the top of the trap and removes ten lobsters, a couple of small *rock crabs*, and four *sea urchins*. Lobsterman Sam doesn't even bother to measure five of the lobsters because his lifelong experience in lobstering has taught him that they were obviously far short of the minimum legal size. He tosses them back into the ocean without even as much as a grimace. But there were five other lobsters that looked to be, without question, of legal size. Grabbing his brass gauge, Lobsterman Sam proceeds to put the gauge to another of the remaining lobsters, but to his surprise it "just misses the gauge" and gets tossed back into the ocean. He proceeds on and puts the gauge to another of the large lobsters and finds that it exceeds the maximum legal size of more than 5 inches, and, with a little reluctance, drops the lobster over the side and back into the ocean. Lobsterman Sam was about to measure the next large lobster but stopped when he noticed that a V-Notch had been cut into the tail section. That lobster was also gently returned to the ocean. There were now just two lobsters left, but it so happened that when he flipped the lobster over onto its backside, the underside of the abdomen was covered with a mass of brown eggs, so that female *berried* lobster was also carefully returned to the ocean.

We now notice that Lobsterman Sam was now down to just one large lobster to be inspected and measured. He reaches out and grabs a hold of the 1½ pounder, the lobster that had to be the one that we had been watching just a few minutes ago when it was in the trap on the bottom. He first inspects the lobster for the presence of either eggs or a V-Notch, and finding neither, measures the animal and determines that it clearly surpasses the minimum legal size. With a little grin on his face, and with the anticipation of a "little jingle" in his pocket, Lobsterman Sam looks around at his *helper*, and then mumbles to the lobster, as if it could really understand, "I finally gotcha!" Omar has met his demise – and was casually placed into the holding tank aboard the boat.

While Lobsterman Sam had been sorting, measuring, and inspecting the yield of the trap, his helper (sternman) had been hard at work removing the old bait and driving a fresh *rack* of fresh bait onto the *bait spindle*. After closing and securing the trap door, he pushes the trap along the gunwale to his helper who, on the signal, pushes it off the *stern* and into the depths of the ocean. After making sure that all was well on the stern, Lobsterman Sam steers his *"Novie"* boat to the next *buoy* where the whole process will be repeated again, and again, and again, until the last for the day will be hauled up, cleaned out, re-baited, and set back into the ocean again!

Those *sub-legal size* lobsters that abound in most lobster traps hauled to the surface by lobstermen are commonly referred to as *"shorts."* Lobstermen, however, might often refer to them by other names, such as *"snappers," "rattlers,"* and *"munchies."* Whatever they are called, they are short of the minimum legal size, and have to be returned to the ocean "until they grow up" and become what lobstermen and lobster marine scientists refer to as *"recruits."* From the time of their young ages, these short lobsters will be hauled up in traps time and time again until one day, when the gauge is put to them, they will be measured and found to satisfy the minimum gauge requirement. They have finally met with the fate of being caught in a trap and removed from the fishery forever by its main predator, the lobster fisherman.

Some lobsters manage to elude the lobster fishermen and never end up in their traps. The size of a particular lobster is most likely the reason, because when it grows to a great proportion it renders itself incapable of entering the trap through the conventional $6\frac{1}{4}$ inch circular opening that leads to the kitchen compartment of the trap. What is to be said of these large lobsters? Where do they go and how do they manage to survive? How large will they become? These questions are often difficult to answer because of the lack of documentation on the subject. While it might be a matter of conjecture, it is thought that many of these large lobsters are great travelers and that they sometimes cover great distances. There is no research available to quantify how long lobsters live once they are large enough to prevent them from entering the lobstermen's traps. Great specimens have been brought up over the years by dragger fishermen, SCUBA divers, and longline fishermen. Several instances of "giant lobsters" caught in the 1800's have been reported by Herrick. He specifically reported on fifteen giant lobsters that had been caught along the coastline of Northeastern North America that ranged from 19 to 34 pounds in weight and 19 to 22 inches in length. Since these monstrosities could not have possibly entered the traps of the inshore lobster fishermen, and most likely not even the much larger *"bear traps"* used extensively by the offshore lobster fishermen, it can be deduced that these lobsters had most likely been brought to the surface by commercial fishermen dragging their weighted nets along the bottom of the ocean, by longline fishermen, and by the SCUBA diver who just happened to be working the bottom and grabbed onto one of these large lobsters that had not traveled far out to sea.

There was the case of a 25 pound lobster caught by a *"longline fisherman"* twenty miles off Chatham, Massachusetts in the winter of 1981 by the Hilliers of Rockport, Massachusetts. This lobster was sold to the American Co-Op in Saugus, Massachusetts for the price of the $110.00 and was subsequently flown to Atlanta, Georgia where it was bought by Halpern's Gourmet Grocery in Atlanta. It seems that the people in Atlanta befriended the beautiful crustacean. Howard Alpern said, "the whole town fell in love with him!" He was petted, given all sorts of nicknames, fussed over, and fed shrimp by hand. The people finally called him "Barnacle Bill" because of his age and the clusters of barnacles that were attached to his old shell.

Phyllis Schwab, an employee of Halpern's, said of Barnacle Bill, "we decided he'd graced the world too long to kill him. We all became quite attached to that animal. He was a great treat." As the story continues to unfold, it turns out that the Halperns had Barnacle Bill flown back to Boston and transported over the road to Saugus, whereupon he was eventually turned over to Pingree and Fred Hillier. And on one fine New England morning, the father and son team who had originally caught Barnacle Bill, took him out into the ocean and released him at a point off Rockport where they thought he would be safe from fishing dragger and SCUBA divers. At the time, Fred Hillier said, "I figure something that's been free for 100 years shouldn't be caught and cooked." Such is the saga of Barnacle Bill the lobster, who hasn't been seen nor heard of ever since!

Such was not the case for "Ralph," a 38 pound lobster that was pulled up in a Virginia scallop tow working the Cultivator Shoals 150 miles east of Gloucester, Massachusetts during the summer of 1982. The 43 inch crustacean, which was dubbed as "Ralph," had lost most of its vigor while aboard the scalloper. His captor and owner, John Nicholson of Easton, Massachusetts refused an offer of $1,000 from a local restaurateur who wanted Ralph mounted and hung from the wall of his restaurant. Another offer came in from the New England Aquarium in Boston. They wanted to buy Ralph for $250, a mere pittance, provided that he was in a vigorous condition. But Ralph was not in all that vigorous condition at that time. Having refused another offer of $1,000, Nicholson decided to have a Stoneham, Massachusetts taxidermist prepare the lobster for mounting. "Ralph" was proudly put on display on Nicholson's dining room wall. He declared that he didn't want to sell the lobster after all, saying, "I wanted Ralph for a keepsake."

For Mark Guisto of Beverly, Massachusetts, a large lobster was caught by yet another method in July of 1978. Guisto grabbed hold of and captured a 15 pound lobster while diving in sixty feet of water off Rockport, Massachusetts, a favorite spot of SCUBA divers who practice their skills there, and in the process, often come home with a basketful of lobsters.

In 1977, the largest lobster supposedly ever caught and <u>preserved for public viewing</u> was put on display at the Museum of Science in Boston, Massachusetts. This incredibly large lobster weighed in at about 42½ pounds and was nearly 3½ feet long, as measured from the tip of the claws to the posterior of the tailfan. It was a male lobster caught in the fall of 1934 by a deep trawl net in 500 feet of water off the Virginia Cape. The lobster was donated to the museum by the General Seafoods Corporation of Milford, Connecticut, but upon investigation at the museum its whereabouts could not be determined. This lobster, however, does not appear to be a record-setter in terms of weight and size. According to the 1984 Book of World Records, a lobster weighing 44 pounds 6 ounces and measuring 3 feet 6 inches in length was caught on February 11, 1977 off the Scotian Shelf, preserved for viewing, and sold to a restaurant in New York City, but the trail of this extremely large lobster has apparently gone cold.

The New England Aquarium in Boston did, at one time, boast about its ownership of the largest <u>living</u> lobster being held in captivity, but that lobster was in poor condition for a period of time and finally succumbed to some form of shell disease. That lobster was a 38 pounder that stretched out to almost 3½ feet in total length.

A visit to such an aquarium offers the viewer an excellent opportunity to observe many types of lobsters on display, including many of the American lobster species that feature several variations in color, such as red, blue, and calico. Usually kept in a separate tank might be found several species of the *spiny* lobster (*palinurus*), but these lobsters are no match for the American lobster in form, structure, or beauty. Having no large claws, they often appear to be a little grotesque to the viewer as compared to the American lobster, *Homarus americanus*.

The European lobster (Homarus gammarus) with symmetrical "crusher" and "ripper" claws. Weight: 4 pounds 10 ounces. From Francis H. Herrick, "Natural History of the American Lobster."

Chapter 5. The Lobster Fisherman

It is without a doubt that the most important element in the lobster fishery is the lobster fisherman, the lobster hunter, the lobster producer, the lobster harvester - or whatever title one might decide to associate with this proud and cultural profession that goes back in history nearly 200 hundred years.

Because of some rather unique personality traits, cultural and traditional background, and the type of work he performs, the lobster fisherman is just about the most humble and super-independent type of person that a stranger would ever happen to meet. And to the unwary, the typical lobster fisherman likes to keep his personal business to himself! There is no need to ask a lobsterman how he's doing or how his catch was for the day, because you'll get pretty much of a stock answer, "I'm gettin' a few!" Quite frankly, the lobsterman doesn't ask you how you're doing in your business, and it doesn't set too well with him when some stranger approaches him and asks him how he's doing in his. To him, it's "none of their business!" Most lobster fishermen are, in fact, people of very few words, and go about their business in a very quiet and unassuming manner.

The typical full-time lobster fisherman is by-and-large a self-employed person and independent contractor. And he likes it that way. Indeed, there is no other way he would have it, for he does what he wants, when he wants, and how he wants, and he doesn't like people telling him what to do. He runs his boat, fishes his traps, repairs his boat and gear, and carries forth an endless number of other activities that are all necessary to the financial success of his business.

Anyone that has been exposed to the activities of the waterfront is sure to realize that lobstermen take their work very seriously. And they take their work even more seriously when aboard their lobster boats and in the process of hauling, baiting up, resetting their traps, and trying to keep track of what traps have already been hauled and what traps haven't.

> [Author's Note: The use of the terms lobsterman, lobstermen, he, his, etc., are used throughout for the purpose of simplicity. While the overwhelming majority of those who fish for lobsters are of the male gender, it is important to acknowledge that they are not exclusive in this endeavor; there are indeed many of the female gender that have been at one time, or are still now, active in this business.]

Most lobstermen are so serious-minded when working on their boats that they steadfastly refuse anyone who would like to go out fishing with them. Robert Stewart, a lobster fisherman working out of Lossiemouth, Scotland puts it this way: "I find taking people out with me a little bit distracting, especially when they are getting bored with the trip. Also, I must be in full control of the boat and my passengers, and many people cannot take orders."

Regarding the question as to who's the" boss" on a lobster boat, or any boat for that matter, let there be no mistake about it: The "boss" is the captain! A reminder to this effect was a plaque hanging in a most conspicuous location in the wheelhouse of Beverly lobsterman Ken Martin's lobster boat. The plaque read: "TO MY CREW: Please Be Reasonable And Do It My Way. THE CAPTAIN".

The ocean is the lobsterman's workplace. Unlike the farmer of the land who must substantially invest in livestock, feed, and crops to be produced for the marketplace, the fisherman of the ocean bears no similar investment because his "crop" already exists in the course of natural reproduction. The lobster, as we have seen, is produced in nature's setting on the bottom of the ocean. And, if the animal can survive the *"water column"* and its enemies of the deep, it will in all probability be hauled up one day in a lobsterman's trap and be marketed as a legal size lobster. So, in a sense of the word, the lobster can be looked upon as a commodity *"that is out there"* and capable of being trapped by anyone with a valid license to "fish" for it and who has the wherewithal to do so. While there is no capital involved in creating the product itself, there is a considerable investment, both initial and recurring, that is necessary to *"catch"* the product and getting that product to the marketplace. The summer tourist at dockside is often inclined to muse and marvel over that *wonderful catch of lobsters* being unloaded from the lobsterman's boat, but most likely will not even give a second thought about the costs of the lobsterman in getting those lobsters to that point.

For any individual who might be contemplating embarking on a career of lobstering, the following menu of investments and expenses might well be considered before doing so. And for a young person to embark on such an undertaking without the benefit of a solid credit footing , and without some form of financial assistance from his family, the goal would appear to be virtually impossible. Let us examine some of the major investments and expenses involved in this undertaking. These investments and expenses are not intended to be all-inclusive, but it gives one a fairly good idea of the scope and magnitude of what is involved in the fishing for lobsters:

Boat-Related Investments and Expenses

- Boat
- Boat Fuel
- Boat Maintenance
- Boat Insurance
- Sonar/Radar Equipment
- Fathometer/Depthfinder Equipment
- Hydraulic Trap-Hauler
- Lobster Holding Tank
- Bait Barrels
- Measuring Gauges
- Foul Weather Gear
- Gaffs
- Rubber Band/Pegs
- Mooring
- Lobster "Car"
- Dinghy

Port-Related Investments and Expenses

- Truck
- Truck Fuel
- Truck Maintenance
- Truck Insurance
- Bait
- Dock Fees (if the dock is not self-owned)
- Lobster Shack Fees (if the shack is not self-owned)
- Association Membership Fees (if a member of one or more Lobstermen's Association)
- Association Membership Fees (if a member of a Lobstermen's Cooperative)
- State Lobster License Fees
- State Boat License Fees
- Mooring Fees

Gear-Related Investments and Expenses

- Traps
- Buoys
- Warp (Rope)
- Bricks

The traditional inshore lobster boat can vary in length from 24 to 40 feet, features a high *"bow"* and a low *"stern,"* and will be powered by either a gasoline or diesel engine. Two types of lobster boats are commonplace in the industry. The first is the so-called *"Novie Boat,"* the second is referred to as the *"Jonesport Boat."* In his book about lobstering, Earl L. Doliber of Marblehead, Massachusetts suggests that "Novies are characterized by high bows and sides that sweep slowly back to a low stern, while Jonesport boats are narrower than Novies and the bottoms are shallower and fatter." Doliber considers the Novie Boats as being easier to handle in rough seas and easier to haul from, while the Jonesport Boats are designed to cut through the water with greater speed. The decision by the lobsterman as to what type of boat he prefers to fish from becomes a little bit of a trade-off. If he fishes most or all of his gear a great distance offshore where the weather can be considerably rougher, he'll probably opt for the more seaworthy Novie Boat; however, he'll have to sacrifice the much faster speed of the Jonesport Boat, and thereby taking a longer period of time to get to his traps and to return to port at the end of the day.

Then there are the other types of "lobster boats" that are operated by *part-time* and *non-professional* lobster hunters. Most are relatively small crafts that are navigated either by rowing or by being powered by a small outboard engine. The boats used by these part-timers and non-professionals are called by several names such as *dinghy, punt, pram, dory,* and *skiff*. Some are equipped with a hydraulic trap-hauler, but most are not, and the lobster traps have to be hauled to the surface by hand which, as one might imagine, is a most laborious task.

The job of a *"lobsterman"* appears to be a nice, easy, and sort of a "cream puff" type of job. He is perceived pulling his boat up at dockside in some snug harbor to unload his catch of the day. He either stores his catch away in a *"lobster car"* or into the rear of his pick-up truck. Then the lobsterman and his helper shuffle off and quickly vanish from the scene. The onlooker at this point mistakably believes that the job of a lobsterman must be "a snap" and that there isn't really much to lobster fishing after all. As we shall see, however, this type of conjecture is far from the truth.

The planning of a lobsterman's day at sea very often starts on the preceding day. He has many concerns on his mind and many matters to ponder before going fishing in the morning. The weather: will it be inclement, will there be a dense fog out there, and will there be a strong and gusty wind? The ocean: will there be strong tidal currents and high swells? The bait: is there enough bait on board or will he have to get it in the morning? The fuel: is there enough fuel on board to get out there, haul his traps, and get back into port again? The help: will his *"sternman"* show up, or show up on time to help him haul the traps, re-bait them, maybe relocate them, and reset them into another location? These are just some of the major considerations that pass through the mind of the lobsterman that must be addressed before going out to sea.

The Tale of the Lobster

The lobster fisherman is not a late sleeper on any morning that he is going to fish. He often arises as early as 3:30 to 4 o'clock in the morning, eats a nourishing breakfast, packs a lunch, and proceeds to the dock in an atmosphere of solitude, with the still and quietness of the dawn being broken only by the call of cranky and noisy seagulls or a "good morning" greeting from his sternman. If his boat is not tied up at the dock, he and his sternman will row or paddle out to where it is tied up to at the mooring. The small boat will be secured to the mooring, the engine will be "fired up" and the lobster boat will be detached from the mooring. If the bait for the day isn't on board or if the boat isn't fueled up, they will have to slowly proceed to the bait shack and to the fuel pump on the dock. After carrying outdoing all of these chores, and being satisfied that the engine is running smoothly, the lobsterman and his sternman will "get underway" and edge out of the harbor and toward the expanse of the open sea.

On a pleasant morning, and with the golden cast of a brilliant sunrise, the "heading out" can be both a beautiful and exhilarating experience. The golden glow of the sunrise permeates the horizon and basks the ocean and all around it with an expression of comforting warmth. During this time of the day the air is warm off the land as compared with several hours later in the day when the air is usually cooler off the ocean and the ocean often becomes a little rougher. This is one good reason why most lobster fishermen are early- risers and get out to sea as soon as possible. It is here, at this time, and in this place, that "getting up with the birds" is the thing to do and to acknowledge without any hesitation that *this is the most beautiful time of the day!* To the contrary, however, are those mornings when the weather works against the best intentions of the lobster fisherman: the sky is overcast, the air is raw and cold, and the rain and the spray from the ocean repeatedly slaps up against the faces of the lobsterman and his sternman. The next few hours can be miserable for the lobsterman and his sternman, but unless the weather is really foul, the two will usually put up with it and go about their work the best they can.

If the lobsterman is fishing his traps close to shore, it will take but a few minutes to reach the first trap-haul of the day. If he is fishing his gear offshore, the trip out to the first trap can take as long as one hour of running time, and perhaps much longer if he is fishing traps in much deeper water far out into the ocean. It should be noted, however, that most lobstermen fish their traps so close to shore that land, except in a fog, is never out of sight. It is only a minority of lobster hunters that will venture many miles out into the ocean to pursue their prey during the coldest months of the year. Many lobster fishermen will haul their traps onto their boats a few at a time and bring them ashore for the season and just as soon as the harshest of the cold winter months makes lobster fishing a trying experience. For them, this is the opportune time to repair their trap gear, repaint their buoys, and tend to the repair of something that is very dear to them - their lobster boats.

The lobsterman must have the ability to know at all times where his traps are located if he expects to haul them aboard. One might say this is nonsense, but once out on the ocean that is awash with a profusion of different colored buoys to consider, lobstermen have been known to get "mixed up." Naturally he is going to haul only the traps that are attached to his distinctively-colored buoys, but in the process of hauling and resetting a traps, or traps, it often happens that the boat gets turned one way or another and is not necessarily headed in the direction of the next trap to be hauled. But lobstermen, after many years of setting and hauling their gear, seem to possess an uncanny sense of keeping track of their trap gear. There might be cases when a lobsterman will haul up a trap that he has already hauled, baited, and reset, but such an occurrence would be the exception rather than the rule. On the other hand, lobstermen have been known to "miss" a trap and it doesn't get hauled at all on that particular day. If a lobsterman's traps are <u>not where they should be</u>, this often gives rise to the suspicion that a *"poacher"* has made the rounds, has lifted the traps, removed their contents, and then dropped them in the same general area. As well as might be imagined, all of this would most likely occur under the cover of darkness. But whoa be it to the poacher! As we shall discover, lobstermen usually "find their man" and they have their own ways of dealing with him.

Some state laws restrict the lobsterman by allowing him to fish only single traps in certain areas. In far more cases, however, he is permitted to fish *"trawls,"* or *"strings."* These trawls or strings connect together as many as four traps, oftentimes more, that are all connected one to the other. Some lobstermen, the hardiest of the lot, fish as many as twenty traps per trawl during certain months of the year when they are fishing far out into the ocean in deep water fishing grounds These lobstermen, tough and hardy souls that they are, often spend more than one day at a time in fishing their traps, most of which are known as *"bear traps,"* and their numbers are few as compared with the thousands of lobster fishermen that operate close to the shoreline.

Robert Delano Martin

When a single trap is being fished, only one distinctively colored buoy is required to connect the buoy to the *"warp" (rope) that* is attached to the trap's *"bridle."* But in the case of a string or trawl consisting of several traps, a buoy is connected to the warp leading to the first trap and another buoy is connected to the warp leading to the last trap. Fishing strings or traps of traps is considered to be a much more productive and efficient procedure for fishing for lobsters. And, should one of the two buoys be cut or severed by man or another watercraft, or as a result of a violent storm, the lobsterman can still haul all of his traps from the other buoy, and thereby not losing his entire string of traps.

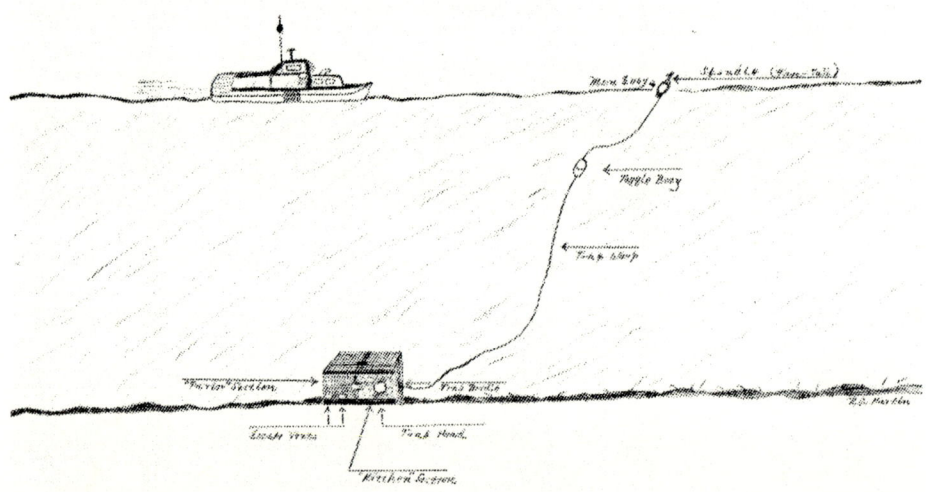

This drawing illustrates a lobster boat heading for a buoy bobbing about on the surface of the ocean. A "toggle buoy" is attached to the trap warp (rope). The trap warp is connected to the "bridle warp" and the bridle warp is attached to the lobster trap itself.

Buoys are usually constructed of a very lightweight Styrofoam material and are painted with a solid color or a combination of two or more colors that are often painted in bands. The color-coding must be unique, may not duplicate the color code design of another lobsterman's buoys, and must be approved for use by the appropriate state licensing authority. Buoys made of materials other than Styrofoam were at one time in common usage. They consisted of contrivances such as wood buoys, milk bottles, bleach bottles, or any other device that would float, but they are not seen anymore. In addition to the color-coding of buoys, state laws usually require that each lobster fisherman paint the same color-coding on both the *"port"* (left) or *"starboard"* (right) sides of the boat's bow - or at the discretion of the lobsterman, one single buoy may be displayed at the top of the highest point over the boat's *"pilothouse."* Virtually all lobstermen prefer this alternative and make it a practice of attaching this single buoy to the tall antenna mast that is associated with their electronic gear.

Many lobstermen prefer a type of buoy that incorporates a *"spindle,"* oftentimes referred to as the *"fan-tail,"* or *"stick."* The spindle, as we shall call it, protrudes from the top of the buoy and renders the buoy much more visible to the lobsterman when he is hauling his traps. A second important aspect of the spindle is that is much more easier to spot during a dense fog than would be the case of just a plain buoy just floating on the surface. A third good reason for the use of a spindle is that it provides the lobsterman with something of real substance to grab onto with one hand as he *"gaffs"* the trap warp with the other hand.

As has been mentioned previously, the distance between the main buoy and the lobster trap setting on the bottom is often referred to as the *"water column."* Many inshore lobstermen employ the use of an additional buoy, called a *"toggle buoy,"* also called a *"bobber,"* that is positioned and secured to the warp a few feet from the buoy on the surface. The toggle buoy, if one is used at all, usually consists of two pieces of circular cork that serves the purpose of keeping the trap warp off the bottom at extremely low tides. The reason for doing so is to reduce the risk of having the trap warp from becoming entangled with other lobster gear, rocks, or any other objects that happen to be lying on the bottom. There are no set guidelines as to where the lobsterman secures a toggle buoy, just as there is no set criteria for the length of warp between the main buoy and the bridle of the warp connected to the trap. Nor is there set criteria for the length of warp between each trap in a string of traps. It is left up to the experience and judgment of the individual lobsterman as to how much *"play"* he wants to make allowance for in the length of warp between these points.

Getting ready to haul, the lobsterman will maneuver his boat to the buoy to be gaffed and hauled aboard. He will come up alongside the buoy, guide the engine a little into reverse, hook the trap warp with his gaff, and then proceed to lift the buoy over the *"gunwale"* of the boat. The buoy is dropped on the deck and kicked to one side while the trap warp is guided over an open-faced pulley, referred to as a *"davit,"* and sometimes called a *"davy."* The trap warp is then inserted into a hydraulic trap-hauler (still called a *"winch"* by some), which will take over all of the mechanical work of hauling the trap warp and the trap itself to the surface. Upon the trap reaching the surface, the strong hands, arms, back, and legs of the lobsterman will work in concert with one another to lift the trap up and onto the gunwale of the boat. From this point onward, the lobsterman and his sternman will coordinate their efforts and functions to maximize the efficiency required in the many operational procedures that will follow.

With the engine now in idle, the lobsterman will unhinge and swing upon the door of the trap and then proceed to remove its contents. What might be found in a lobster trap? Sometimes absolutely nothing! But in most trap hauls, there might be lobsters of various sizes, and in some traps there might be crabs, starfish, sea urchins, fish, or perhaps even a rusty beer can or an old shoe or boot! Most lobstermen do not like the spiny and prickly sea urchin, and in the jargon of some lobstermen, they are referred to as *"whore's eggs."* In recent years, however, some of these same lobstermen have changed their tune about the sea urchin, since a market was found for them in the Far East, notably in Japan.

Even though most lobster fishermen employ a sternman, they usually handle the chore of cleaning out the traps themselves. When a trap contains a lobster of questionable size, he will measure the lobster's *"carapace"* with a brass gauge. All legal size lobsters are called *"keepers"* or *"counters,"* meaning that they may be kept and counted as part of their catch of the day. A legal size lobster will have both its crusher and ripper claws shut tight with heavy rubber bands before being placed into some type of tank or other lobster-holding container. The rubber band used to immobilize a lobster's claw is stretched wide open with a device similar to a pair of scissors and referred to by some as *"band spreader."* While all professional lobstermen prize their catch and keep their lobsters healthy and vigorous by storing them in a holding tank that is kept full with re-circulating seawater, a small number of lobster hunters will simply use a bushel basket for the storing of their lobsters. If the weather is a tad bit on the warm side, the lobsters will be covered with burlap or seaweed and will be periodically doused with fresh seawater. Lobsters will remain alive and in excellent condition if they are protected from bright sunlight and sudden changes in temperature.

Sub-legal size lobsters, generally referred to as *"shorts,"* are immediately returned to the ocean. In Maine and most other lobster-producing states, *"oviferous"* (egg-bearing) female lobsters are also returned to the ocean. The same holds true for any female lobsters that have a *"V-Notch"* cut into one of a blade-like appendage that make up the *"telson"* or *"tailfan"* of the lobster. Also some states have followed a Maine lobster preservation measure by returning to the ocean any lobster with a carapace length of more than 5 inches.

Most lobstermen, without question, handle the sub-legal size lobsters with the utmost of care. However, there will always be a few of their profession - and I have witnessed it firsthand - that mistreat the sub-legal size lobsters and handle them very roughly when removing them from their traps and *throwing* them back into the ocean A lobster that has one or both of its large claws locked in a death grip on the netting of the trap will most likely not yield that grip. It will hang on for dear life, and a lobsterman growing short of patience might attempt to yank the claws loose from the netting. The loss of one or both claws and the attached knuckle appendage(s) will often result. And what a shame, for that lobster has not only become a *"cull lobster"* for a period of time, but its main weapons of defense and attack will have been lost, and it will take two or three *"molts"* or more for that lobster to regenerate these appendages to their original form and size. Some fishing areas, especially close to shore, abound with sea urchins and undesirable fish that go for the bait in the traps. Oftentimes the lobstermen will shake and bounce the trap on the gunwale to free them up. Then he will turn the trap upside down to shake them loose so that they will descend to the bottom. It has been observed that some lobstermen so dislike the creatures that, rather than returning them to the ocean, they will bring them back and destroy them on shore. The fact that most lobstermen *get sick and tired* of continually hauling these creatures up in their traps is without question, and this is especially so when they are having to put up with bad weather, are having a bad day, and the catch is disappointing.

Trap repair is pretty much of an ongoing practice on a lobster boat. Trap repair is usually done right on the spot by either the lobsterman or his sternman, because any large gaping hole in the netting or any large separation in the *"lathing"* of a wooden trap might allow even legal size lobsters to escape. Should the damage be extensive, or if the trap is very old and has seen it's better days, it might be sacrificed to the ocean or brought back to shore and be destroyed.

With the contents of the trap removed, and with all minor repairs having been accomplished, it is customary for the lobsterman or his sternman to remove the old bait and for the sternman to put fresh bait onto the *"bait spindle."* This being done, the lobsterman will close the trap door and push the trap along the gunwale *(washrail/washboard)* to the sternman who will store it more aft on the gunwale or in the *"aft" (rear/stern)* of the boat. It is from either one of these two locations that the trap will be shoved off the boat and into the ocean again.

The type of bait used by the lobster fisherman is often a matter of choice and he usually has personal preference in this regard. On the other hand, he might not have a choice in the matter at all because of the scarcity of bait during certain times of the year. He simply has to use whatever bait is made available by the *"bait man."* Listed on the menu of preferred baits are large herring, pogies (menhaden of the herring family) and some species of *"flatfish"* such as flounder or sole. But probably the most favored bait of all is the redfish, more properly named ocean perch, and often referred to by lobstermen as *"brim."* The next most popular bait in the business seems to be the skeleton remains of flounder and sole which are usually *"bagged"* rather than *"racked."* In lieu of employing just the use of a bait spindle to accommodate a rack of fish, such as redfish, lobstermen use a nylon mesh bag to keep the bait confined to the bait spindle area of the trap. The bait spindle is elevated in the *"kitchen"* section of the trap and close to the hooped opening that leads to the *"parlor"* section of the trap.

At times when the supply of bait is especially scarce and expensive, the lobster fisherman will often resort to the use of large crabs for his bait. These crabs, either bought from his bait man or hauled up in his traps, will be intentionally smashed and crushed to immobilize them and allow their juices and body fluids to be dispersed in and around the trap and thereby making it tempting for lobsters to enter the trap. Any small species of groundfish that manage to find their way into the kitchen compartment of the trap will be treated in a similar manner in order to lure lobsters into the trap.

On the list of common nuisances to the bait in lobster traps is the *"sand flea"* which is only about 1/16 of an inch in length. Also on the list are *"starfish,"* *"snails,"* and *"krills,"* the latter of which measures about a half inch in length and resembles a flea. These forms of marine life are pests to the lobstermen in the sense that they are very skilled in devouring his bait. And then there is the *"skulpin,"* a rather ugly type of bottom fish that is capable of entering the trap and picking the bait *"clean as a whistle."* If a skulpin is pulled up inside a trap, however, it is usually dealt a staggering blow and tossed into the trap for lobster bait. For those relatively few lobstermen who "hate to give up the ship" and are still fishing wood traps, there is a marine organism, the *worm borer,"* that will bore through the tiniest opening in the wood lathing, and once inside the lathing it will continue to bore away and in time will destroy the trap.

After hauling a single trap *(singles),* two traps connected to each other *(doubles),* three traps in a trawl *(triples),* or many more traps in a trawl, one of the foremost thoughts in the lobsterman's mind is the consideration as to whether to reset his traps in the same location or to relocate them elsewhere. If the catch is up to par where he is, the chances are good that he will reset them in the same area; if the catch is not up to his expectations, he will no doubt decide to go some distance to *"try out a new spot."* - perhaps a spot where there are other fishermen working their gear. This is often a telltale sign that there are lobsters there and that the catch must have pretty good for those lobsterman. On the other hand, an area might be strewn with scores of lobster buoys floating on the surface and with lobstermen actively setting and hauling their traps. The lobsterman might then ask: "Should I consign myself to the fact that the area is just *too busy?"* and "Wouldn't it be better for me to move on?" He would therefore be adopting the mind-set of long-time lobsterman Herbert Baum, Jr. of Kennebunk, Maine, who remarked,

> "The secret of bein' a good lobsterman is getting where somebody hasn't been and knowin' where they're gonna' be next.
> "So you try to put them (traps) where somebody else isn't and hope your guess is betta' than theirs."

The lobsterman and his sternman will go through the same rudimentary motions, trap after trap, until they have hauled as many of them as is physically possible, or that they want to haul for that day. The number of traps to be hauled on any given day will be dependent on a number of factors, but if he has enough bait on board, the weather is agreeable, and things are going along fairly well, he and his sternman will probably haul all the traps - unless they have put out so many traps that hauling them all in a single day would be a virtual impossibility. And there are many, if not most, lobstermen that fall into this category. They have built or purchased and set out so many traps that they are forced to let a good number of them *"set-over"* for a day or two between trap hauls. These lobstermen will haul a portion of their traps one day, another portion of them on the next day out, and so forth until all of the traps have been hauled, re-baited, and set back down into the ocean again. In the lobster business, any day that a trap is not hauled is referred to as a *"set-over-day."* Set-over-days might best be described by using the example of a lobsterman who is fishing 800 traps. Chances are that after going to this effort and expense, he is going to fish those 800 traps *"come hell or high water"* rather than leaving a good portion of them to clutter us the dock, his shack, or even his back yard. Before getting his traps into the water in the spring, he naturally realizes that he can't possibly haul and reset all of them in a day's time. So perhaps he will haul 300 of them on Monday, another 250 traps on Tuesday, and the remaining 250 traps on Wednesday, all of which will be contingent on the weather, the bait supply, engine trouble, the availability of his sternman, and even personal sickness of one form or another. Then he'll repeat pretty much the same process on Thursday, Friday, and Saturday. At least Sunday would be a *"day off"* for the Maine lobsterman during June, July, and August because fishing for lobsters on Sundays is prohibited at this time of the year.

A survey of Maine lobstermen was conducted which clearly demonstrated both the extensive and intensive lobster fishing that goes on in Maine waters. Of the lobstermen who responded, 4,150 of them said that they fished 350 traps or less, 800 of them fished between 450 and 600 traps, and 100 of them fished 750 traps or more. However, two notes of caution are appropriate at this point: the first is that the number of responses returned represented only a portion of the total number of survey forms sent out; secondly, many Maine lobstermen are known to be on the conservative side with their numbers. In this sense, therefore, the number of traps reported as being fished is probably understated.

Robert Delano Martin

A typical late afternoon scene at Boothbay Harbor, Maine with a lobster boat at rest after a day of fishing. This photograph illustrates the manner in which this particular lobster fisherman maintains his boat in an immaculate condition. The small boat tied up alongside is this lobsterman's way to get out to his mooring and back. In the background is another type of lobster boat, much smaller in size, and which is typical of the boats used by "part-timers" or other lobster hunters who have fewer traps set out in the ocean. In this case, there would be less hauling and setting of traps, thereby making it unnecessary to use one of the much larger conventional lobster boats.

The Tale of the Lobster

When heading back to port after a long day of *"hauling and setting,"* the sternman will undertake his customary job of swabbing down the boat and getting it *"ship-shape"* for the next day of fishing. Most lobstermen take a great deal of pride in their boats and are very particular in the manner of its cleanliness. And it is common knowledge that an unclean and smelly lobster boat not only attracts seagulls, but these seagulls often leave their *"calling cards"* as well.

The lobster catch for the day is usually sorted out by weight, size, and condition. With regard to the latter, *"cull lobsters"* are usually separated from the more perfect two-claw lobsters. A cull lobster, it may be recalled, is a lobster with only one large claw - either the *"crusher claw"* or the *"ripper claw"* - or no large claws at all. Cull lobsters customarily yield a lower price for the lobsterman. Some lobstermen have, however, built up a trusting relationship with their dealers over the years and do not have to meet the requirement that the cull lobsters have be *"culled"* out of the total catch. This being the case, the lobsters are weighed and dumped as a lot into one or more of the dealer's crates in a floating device called a *"lobster car,"* also called a *"live car."* If the lobsters are to be delivered to an establishment that sells directly to the public, then chances are that they will have to be culled out. If the lobsterman decides "to sit" on one or more days of lobster landings, he will store them in crates which will be dumped into his "lobster car." The crates are similar in design to the lobster trap and hold about 100 pounds of live lobsters. The lobsters will remain there until such time as "the price is right" or the occasion occurs when the lobsterman simply has to sell them in order to pay his bills.

Lobstermen who fish a sizable number of traps will usually employ a sternman in order to get the job done and to make his work easier. Edward C. Blackmore of Stonington, Maine, a lobsterman and a former longtime president of the Maine Lobstermen's Association, stated his personal reason for having a sternman aboard: "I've been fishing for a long time and I'd rather be working the stern than running the boat. There's a lot more work to it in one sense of the word. I don't want his pay, no, but as far as the work goes, I'd rather have his job." Then there are the concerns about safety on the ocean. During one of the annual Fishermen's Forums at Rockland, Maine, one lobsterman lamented: "In my case, I wouldn't go out of the harbor alone right now. I've had too many things happen and I wouldn't get caught out in that boat alone. It isn't worth it!"

Robert Stewart, a lobsterman from Lossiemouth, Scotland, had this to say about his personal experiences as a professional lobsterman:

> "Admiring glances from the quayside may be very uplifting for your ego, but there will be nobody to admire your courage outside the harbour when it becomes the turn of your stomach to become uplifted.
> "The life of a lobsterman can be very hard when the weather is miserable and cold and his pots are consistently empty. On the credit size, of course, are the good times. Times when no other life could be considered as an alternative, for anything else would seem empty and routine by comparison.
> "For make no mistake about it. It is and must be considered a business if you want to make a success of it. It's no good sitting and daydreaming on a nice summer's day looking out over a glassy sea and thinking how nice it would be to get away from that stuffy office and simply make money while enjoying yourself "messing around in boats." Consider the hard times when you return wet and bedraggled with very little to show for a hard day's work in icy conditions with the sea behaving in a most unlikable manner.

"Taken into account the long hours you will have to spend repairing and maintaining your boat and gear. Then if you still have a spark of adventure left, is the time to look for the brighter side. You are your own boss. Whatever you put into the business is for your own benefit.

"Any type of fishing is of course a gamble, and the thing to remember is that you will get out of it what you are prepared to put into it. So, once having made up your mind to take the plunge, be prepared to experiment, test new equipment and ideas and generally consider new ways in which your methods can be improved. I have found this pays off reasonable well, and I think I have proved my points in these notes.

"It is just possible that stocks aren't as good as "in the good old days," but there are often pleasant surprises in this business."

So one might ask, "but how easy or difficult is it going to be to go fishing for lobsters?" It's not that easy at all! In Canada, for example, various lobster management plans are in place to maintain the economic viability of the lobster fishery and to ensure that the fishery does not become depleted. And in the two major lobster-producing states of Maine and Massachusetts, there are all sorts of possible restraints to enter the fishery, such as minimum age, residency requirements, the number of traps that can be fished, etc. These and other restraints regarding the lobster fishery are examined in Chapter 6 where an emphasis is especially focused on the Maine lobster fishery.

Anyone who has an inkling about *"settling out a few traps"* might perhaps give the matter a second thought. This is especially true in Maine waters where, because of its cultural and family tradition, such a practice would not be taken too lightly by many lobstermen who make their living by fishing for lobsters. Edward C. Blackmore had some personal thoughts on this subject:

"In Portland, they tell me, anyone can fish, but you couldn't do that here in Stonington. You wouldn't have half your gear left. You have to be known, you have to have some connection to the industry... have been a sternman for someone else. You have to have acceptance from the fishermen or they'll pick you to death, one to ten traps a day. Even a native might have to serve an initiation. Someone might lean on him a little bit, an initiation, a token, not too drive him out but to show he cannot drive anyone else out; that he has to fit his gear in with the gear of others.

"There are certain areas right here in Stonington that I can't fish now. Would cut my gear, and I've been fishing here for 30 years, and I'm a native-native. It's our livelihood. If somebody hurts the industry, there's no place to go. You got to sell your home and go."

Lobsterman Robert Stewart of Scotland pretty much states the same chapter and verse: "There is no system of staking a claim to an area and legal method of keeping strangers off your favourite ground, but newcomers have been know to suffer unusual losses and have been finally discouraged."

In a survey conducted by the author a few years ago, the following question was asked: "In your opinion, how do professional lobstermen look upon "outsiders" setting traps in "their" fishing areas?" The following responses from some of these people who have knowledge of the entire lobster fishing industry are indicative that *"territorial rights,"* although not sanctioned by law, are indeed a very real circumstance - and are usually enforced - especially in waters off the Canadian Maritime Provinces and *"Downeast Maine."*

"Not very favorably, and with justification under present circumstances. However, territoriality is probably declining in the lobster fishery as a result of technological change that has occurred over the last twenty years."

"Professional lobstermen resent them (outsiders) and have their own ways of dealing with outsiders."

> "In some places they look upon them (outsiders) through the sights of a rifle. Lobstermen are more territorial than their prey, and their fishing is, in most places, closed to outsiders."
> "Object!"
> "With anger!"
> "They don't usually react badly if the new man 'is a good guy'."

Who might be considered as an "outsider" when it comes to setting out a few traps? He or she would generally be a person residing beyond the fishing town's boundary lines, and even more so in the case of a summer vacationer or "city person" who just wants to come up and throw a few traps into the water. And it doesn't seem to matter even if the person is a "Downeast Maine" resident. It appears that the further "Downeast Maine" one travels, the stronger the so-called "territorial rights" will prevail, and that these rights are in place not because of any state law, but because of very ingrained social, cultural, and family traditions that have manifested themselves over the years.

[Author's Note: The term *"Downeast Maine"* is rather confusing, as one usually considers "down" as being in a rather southerly direction. However, in the minds of most Maine residents, the term "down" really means "up" and being in a rather northerly or northeasterly direction.]

When confronted with the "part-time" lobstering issue, many lobstermen and observers are quick to draw from the hip to foster their views on the subject. In *"The Salt Book,"* author Marshall Alexander of Bradford Pool, Maine, remarks: "I am death against part-time fishermen. It goes back to whether the lobster business is going to be an industry or whether it's going to be a sport." When Kenneth Martin of Beverly, Massachusetts was asked if he would favor restricting the fishing for lobsters to full-time lobstermen, he replied: "Yes. Definitely. MDs wouldn't allow part-times to practice medicine." And in his book, *"Lobstering – Inshore and Offshore,"* Earl L. Doliber of Marblehead, Massachusetts makes his feelings known as to how he feels about the part-timer issue. He refers to them as *"rag pickers"* and makes the inference that they are resented and mistrusted by full-time lobstermen in the area.

It is a generally wide-held position that not too much resistance would be given to someone just starting out in the lobstering business if that person was a local lad in school who wants to set out a few traps and who has made his intentions known that he intends to "go lobstering" after completing school - provided, of course, that "he keeps his nose clean" and abides by the general rules of the professional lobstermen - and is regarded as a well-liked youngster. Such, however, would probably not be true for an adult who is holding a regular full-time job and wants to fish for lobsters "on the side." He would not fare the same level of acceptance that Edward C. Blackmore talked about, and whether he is a "townie or not, he would most likely be looked upon as an interloper, wanting to derive an income from two sources, and cutting into the financial rewards of the full-time lobstermen whose sole income is derived from fishing for lobsters.

Having discussed the generalities of outsiders and part-time vs. full-time lobstering, there is need to touch upon two areas of concern to the lobsterman when he is in the process of operating and managing his business. Probably the foremost in his mind in the concern about the "weather." It should really come as no surprise that adverse weather conditions are among the main reasons for lobstering to be a largely seasonal business. The harsh and cold winter months bid no safe abode for the lobsterman and especially for those that work out of smaller boats. Fishing for lobsters during these months cannot only be dangerous, but the high loss of trap gear works against the bottom line of his business. Even during the months when more favorable weather conditions prevail, the toiler of the ocean will always keep a cautious eye to the sky, the wind, and the swells of the ocean. A rough ocean with harsh winds and deep ocean swells will keep most lobster boats in port. And so will a "pea soup fog!" While a thick fog is often accompanied by a calm and glassy ocean, navigation through it can be extremely dangerous. If he does put out to sea in a thick fog, the lobsterman not only puts himself in danger, but he is hampered by not being able to readily spot his buoys. This amounts to an increase in effort, a decrease in efficiency, and the longer amount of time he will have to expend in hauling and resetting his trap gear. The lobster fishermen, therefore, will usually bide his time and wait patiently for the fog to burn off before heading out to his fishing grounds. Or, if he lacks the patience on that particular day, he might just put fishing on the back burner and not go out on that day at all.

Then there are those *"natural landmarks"* on the ocean that all boaters have to pay strict attention to while navigating through a thick fog and when the tide is ebbing and flowing. Along the New England coastline, especially in Maine, New Hampshire, and Massachusetts, there are small islands, outcropping of ledge, navigational buoys, and warning devices that are a threat while a thick fog might be masking their presence. Outcroppings of ledge that might be of little concern and consequence at high tide can become a real peril at low tide or in between high tide and low tide. To complicate matters further would be the presence of a profusion of lobster trap buoys bobbing about on the surface and often difficult to see because of these foggy conditions. The *"skipper"* of any watercraft must be alert to the presence of these buoys, because a boat's sharp propeller can very quickly sever a buoy from its warp that leads to the lobster trap - and if that trap is not connected to a string or trawl of traps - it will be lost forever and serve only the purpose of resting on the bottom as a *"ghost trap."* This being bad enough, the warp leading to a trap might get wrapped several times around the boat's propeller, causing havoc and taking quite some time to free the trap warp loose.

Although the exact number is not known, it is estimated that far less than one-half of the lobster fishermen fish for lobsters throughout the year, while the remainder fish on a seasonal basis. There are many professional lobstermen that fish for lobsters only during certain months of the year and then switch over to fishing for other shellfish such as shrimp and scallops during some of the other months of the year. While they might prefer to stay with lobstering, they realize that in addition to bad weather, many lobsters have the tendency to go off their food and seek seclusion during the coldest months. It is also known as fact that certain numbers of the lobster population will avoid the colder seawater temperatures along the coast and migrate seaward and into greater water depths where the seawater temperatures are warmer. It may be concluded, therefore, that there are many factors that influence the lobsterman to stay in port and remove himself from the lobster business. Having said this, all is not lost, because some of the lobsterman's time has to be set aside for the repair and maintenance of his boat, his traps, and other associated fishing gear. Traps and gear (warp, buoys, and the like) are removed from the ocean a few at a time, brought to shore, and worked upon during these cold winter months Once upon shore, all of the gear is then muscled in and out of the lobsterman's pick-up truck. Those traps that do not need attention will be stacked in the lobsterman's storage space on or near the dock; Some lobstermen are fortunate to have a handy place to work on their traps and gear, such as a *"lobster shack,"* while others might use the basements of their homes to carry out this work. This will be the time, and probably the only opportune time, for the boat to be hauled out of the water to rid itself of barnacles, mussels, seaweed, and other marine life that has to be scraped from the hull prior to application of one or more coats of primer and marine finish paint.

After viewing the entire landscape pertaining to lobsters and the lobster fishery in general, it is reasonable for one to draw the perspective that the most singular important attribute to the lobster fishery is the lobsterman himself. He is, without question, the apex, the catalyst, and the driving force of the industry. Every practice and procedure is subordinate to him, for it is he that toils long hours at sea, hauls and sets hundreds of traps a day, and tries to outguess the lobsters and other lobstermen as to where they are and where they are not. It is he the lobster fisherman who puts up with the rain, the raw wind, the fog, and the swell of the ocean. He is a man of the sea who has to contend with the bad days when the catch is off and wasn't worth going after at all. It is he, the lobster fisherman, who stands for hours on end while steering his boat, hauling and setting his traps, and having to put up with the stench of the bait barrels. Yet he is no whiner, and despite the contribution of the other elements of the business that bring the lobster to restaurants and the dinner table, the lobsterman is the harvester and producer that makes it all possible.

In the previous chapter, *"Life On The Bottom,"* there was a focus on how the lobster carries out its life on the bottom, how it behaves when entering a trap in search for food, how a legal size lobster attempts to escape from the trap, and how that lobster meets its demise in trying to do so. In the current chapter, *"The Lobster Fisherman,"* we shall expend our attention to the lobster fisherman, how he fishes, where and when he fishes, and sundry other functions that are ongoing both on and off the lobster boat. Drawing now to the conclusion of this chapter, a concentrated focus will be placed on ***one*** Maine lobsterman who, in the author's opinion, can be considered as being typical of most, if not all, of the professional lobstermen who ply their boats along the coastal waterways in search of *"Homarus americanus,"* the American lobster.

Over the years I have *"gone lobstering"* with fishermen from Scituate, Massachusetts to Stonington, Maine. They were all truly fine gentlemen. A few of them, however, would not voluntarily open up to me; it was like they were trying to keep secrets from me or hide something from me when confidentiality was assured. So some time ago I telephoned Herbert Baum, Jr. of Kennebunkport, Maine and expressed my desire to go out fishing with him. He replied that he was just recuperating from an illness. After expressing my sympathy with his situation, I asked him what other lobsterman, of all his lobstermen acquaintances, would be a good person to contact. Such a man, I said, should preferably be a person who is conversant, and that he must be a Maine lobsterman. After some deliberation and soul-searching, Herb Baum offered up two names and then proceeded to narrow his choice to just one. That name was:

Arnold (Joe) Nickerson

Of

Kennebunkport, Maine

I managed to reach Arnold (Joe) Nickerson at his home in Kennebunkport. After telling me that it was *"thick – a – fog out thar"* and that he hadn't gone out fishing that day, he very willingly accepted my request to spend the following day with him while he hauled and set his traps. He informed me that he fished out of Cape Porpoise Harbor, had a new sternman (in this case a sternwoman), and that he would probably be moving a lot of gear around because of the threat of a *"nor'easter,"* possibly even a hurricane. Joe Nickerson, as he prefers to be called, suggested that I meet at the town fish pier as late as 6:30AM because he and his sternman had to get the boat off the mooring and get bait loaded onboard. His final suggestion was that I bring along a lunch and that I wear a pair of rubber boots because his boat had a low *"scupper."*

Having risen at 4AM, I proceeded up Route 95 in Massachusetts and along the New Hampshire and Maine Turnpikes, then east through Kennebunk, Kennebunkport, and into the little town of Cape Porpoise. I nervously arrived at the pier a few minutes late because I had blown right through the intersection in the middle of town that he had told me was *"the square."* That was some square! But, on the other hand, that was a square in the tiny seaport in Cape Porpoise, Maine - and not the much larger square that one might find in my hometown city of Beverly, Massachusetts.

"I'm looking for Joe Nickerson" was my greeting to two lobstermen who were loading crates of live lobsters onto a truck. One of them responded, "He's down at the end of the pier havin' engin' trouble. Hadd'a be towed in from the moorin'." I mused to myself, "oh no, not after getting up so early and driving all this way!" But as good fortune would have it, I met up with Joe Nickerson and his helper, the engine turned over, and after a few pleasantries, I boarded the boat. Within a few minutes or so, two barrels of bait were lowered into the boat and we proceeded to head slowly out of Cape Porpoise Harbor. It was about 7:15AM, actually a late departure for a lobsterman, on a calm and sunny September morning as we cleared the inner harbor and began to maneuver through a channel so profuse with colorful lobster buoys that it was like trying to navigate through a minefield.

Because lobsterman Joe Nickerson fishes most of his traps close to shore, we were able to reach the first of his traps in just a manner of minutes. He gentled the engine, reached for his gaff, and then swooped the gaff under the warp attached to the buoy. The buoy, marked "NICK 3875," along with a toggle buoy that followed, were thrown onto the deck in the cockpit. The warp was then run through an open-faced pulley and into the pick-up of the hydraulic trap-hauler.

The Tale of the Lobster

Joe Nickerson's buoys were painted a colorful red and yellow and each of them had a wooden *"fan-tail"* that was also color-branded with red and yellow paint. The "NICK 3875" was also emblazoned on each of Joe Nickerson's lobster crates and bait barrels, and all of his lobster traps bore a unique tag used for the purpose of identifying ownership.

The first trap to be hauled turned out to be one of the better *"trap hauls"* of the day. It might have been considered as a good omen because it yielded four *"keepers."* It was one of the scores of traps hauled that day that were built by Joe Nickerson himself. At the time, he was fishing about 800 steel/vinyl-coated wire traps, each of which measured about 4 feet long, 2 feet wide, and 1 foot high. They had three compartments rather than the conventional two compartment traps used by a good many of Maine's lobster fishermen who fish the coastal inshore waters. The Nickerson three-compartment trap features a *"kitchen"* compartment where the bait is stored, and two *"parlor"* compartments where most of the contents of the trap usually end up. Each of his traps contained three bricks for ballast and heavy oak *"runners"* on the bottom to keep the trap steady and making it less apt to shift and tumble during times of severe turbulence on the bottom of the ocean.

Each trap contained at least two *"escape vents"* to allow for the escape of sub-legal size lobsters *("shorts"), while* at the same time preventing legal size lobsters *("keepers")* from escaping the trap. The inclusion of escape vents, in this case of rectangular design, is the law in lobster-producing states as a conservation measure to reduce the rate of injury and mortality of sub-legal size lobsters that are continually hauled up in traps, handled time and time again, and then tossed back into the ocean time and time again. Nickerson's traps, like most of the traps of other lobstermen, are equipped with square-shaped *"bio-degradable panels"* that are located on the top of the parlor compartments and which will more or less self-destruct over time to allow lobsters to escape the trap should the warp between the trap's bridle and the buoy on the surface be severed. Such an occurrence would result in the trap staying on the bottom, most likely for years, and become what is known as a *"ghost trap."*

As lobsterman Joe Nickerson was busily engaged in cleaning out the trap of its contents, his sternman was set busy taking bait from the bait barrel and stuffing it into a nylon mesh bag known as the *"bait spindle."* The bait of the day was medium size herring, which was ripe, but not putrid. Having cleared the trap and having the trap re-baited, he pushes the trap along the gunwale towards the stern of the boat and into the hands of the sternman. The sternman then returns to the cockpit in order to retrieve the buoy, the warp, and the toggle buoy. With the main buoy in one hand, he then *"boot-slides"* the warp and toggle buoy along the deck to the aft of the boat where they are positioned alongside their associated buoy.

Lobsterman Joe Nickerson usually fishes all *"singles,"* although *"doubles,"* *"triples,"* or even several more traps in a trawl are permitted within the *"three-mile limit"* off York County, Maine. His traps are usually shuffled around from one location to another, but he does so only if he is working on a hunch that the lobsters are not there in sufficient numbers to make it worthwhile to continue fishing in that particular spot. Once having made the decision to relocate several traps, the work of the sternman takes on additional functions. He now becomes *"busy-busy,"* because in addition to his duties previously described, he must now stack and position the heavy traps on top of the gunwale or in the stern of the boat. Then when the skipper has determined what he judges to be a *"good spot"* for his traps to be reset, they are then shoved over the side or the stern one-at-a-time and into the ocean.

In the process of determining a new location for resetting his traps, Joe Nickerson keeps a casual eye on the lobstering activity going on about him and a very steady eye on his *"fathometer/depthfinder."* His electronic gear is his technological right arm, which will provide an accurate indication of how deep the water is, and a fairly good impression of the make-up of the ocean bottom. He often goes round and around in a particular area until he is confident that he has found a spot where there is likely to be lobsters. He explains it this way:

> "Inshore if you're fishing the hard bottom you try to set your traps over the edges of ledges or between hard rocks - as that seems to be where the lobsters stay. If you're fishing gullies, you try to fish the deep spots, as that seems to be the best spot to be. Sometimes when fishing offshore it makes no difference where you set - the lobsters are everywhere.
> "After you set your trap, and while you keep glancing at your depth recorder, you might see a deep spot that looks good to set your trap in.
> "After doing this over the years, and taking landmarks and Loran fixes, you get to know the bottom, and that gives you an edge over the newcomer in the business."

When ready to reset a trap, Joe Nickerson calmly, almost sheepishly, says to his sternman: "Yup!" or "Go Ahead!" or "Okay!" With the main and toggle buoys in one hand, the sternman pushes the trap off into the ocean. When almost the entire warp has been *"played out,"* the sternman throws the buoys into the water, and within a few seconds all that remains in sight is that red and yellow buoy bobbing about on the surface of the ocean.

On this particular day of fishing there was a group of lobster traps that lobsterman Nickerson was contemplating moving to another location. Then what happened changed his mind: Up came a trap with three *"keepers,"* and then the next trap with two *"keepers."* He turned to me, and with a little grin on his face, said, "I can't see running away from lobsters!" Needless to say, he did not relocate those traps and he reset them right back to where they came from.

Joe Nickerson's attitude and demeanor are very similar to other lobstermen who have allowed me on their boats. Standing in the *"pilothouse,"* he has one hand on the wheel and the other hand on the throttle lever. He will periodically look back to the stern to put a check on how well his sternman is doing. He is cautious about everything that is going on round about him while in the process of lining up for the next trap haul and while hauling and setting his gear. The safety of the crew and the boat is his major concern because he has learned from personal experiences, and from the experiences of others, that lobster fishing can be a dangerous and serious business.

The dialogue between lobsterman and sternman is confined to only the essentials; there is no meaningless conversation that could bring about a distraction from the business at hand. And this is justifiable because the hauling, the cleaning out, the re-baiting, and the resetting of hundreds of traps is a very fast-paced operation that demands concentration and coordination of both the lobsterman and his sternman. On any lobster boat, there is no quarter for any activity than can detract and take the mind-set away from the multiplicity of ongoing work functions, as these functions must be carried out in and efficient and safe manner.

When it is necessary to stack a grouping of lobster traps along the gunwale or on the stern deck, their storage along with their associated warp and buoys can create one of the biggest problems for the sternman. And the lobsterman knows that! On this day of fishing there were several occasions when the sternman had a dozen or more traps stacked aft and ready to be pushed off one at a time when the signal was given to do so. The pushing-off, the playing out of the warp, and the tossing of the buoys of each trap was being carefully watched by Joe Nickerson. Yet, despite the generally good job being performed by the sternman, there were three traps and associated gear that became entangled with one another and went off the stern at the same time. Requiring quick and drastic action on the part of Joe Nickerson, he immediately threw the engine into reverse, then into idle, whereupon he rushed aft to help his sternman who was trying to hang on and attempt to untangle the mess. And during the process of doing so, both the skipper and sternman had to be ever mindful of the danger of getting some of the trapped warp wrapped around their rubber boots, which conceivably result in being pulled overboard and into the ocean.

This fine September day was one of the better days of fishing for lobsterman Joe Nickerson. But make no mistake about it - all days do not produce the same results - such as the day's catch on a cold April day when I went out fishing with Ed Blackmore of Stonington, Maine. Ed caught three legal size lobsters and a bushel of crabs!

[Author's Note: The following discourse relates to some of the more important observations that were made while out fishing for lobsters off Cape Porpoise, Maine. This is just about as close as one might ever expect to get to such an event and to a real-life Maine lobsterman, to some of his personal philosophies concerning the business, and to some of the insights into the happenings while fishing for lobsters in Maine coastal waters.

I have here and there throughout this chapter taken literary privilege to project some personal insights with respect to the subject matter - all of which I hold strong beliefs - and which I sincerely hope will provide the reader with a better, a more true, and a more interesting description of the lobster fishery during the latter part of the 20th Century

If some of the material in a few instances might appear to be somewhat repetitive in nature, this is quite intentional on my part in order to reaffirm certain important points. I trust that the subject matter contained in this writing will be beneficial to both casual and academic readers alike who are interested in knowing and learning about the American lobster and the lobster fishery]

A "Keeper" Or Not A "Keeper"?

On this particular outing, lobsters averaging from about 1 to 1-1/8 pounds were common and with the majority of them weighing in at just about one pound. It was interesting to note that a few of these legal size lobsters didn't even tip the scale at one pound! However, we must remember that lobsters are measured not by weight but by the length of the *"carapace."* Onboard a lobster boat the question of legal size is usually a mute question since experience over the years should have afforded the lobsterman with the ability of knowing without having to measure it with a gauge. Lobsterman Joe Nickerson has over the years acquired this ability, and as a result, many of the lobsters he handled were immediately placed into the holding tank without him having to bother going for the gauge. The same held true for the majority of lobsters that were obviously of sub-legal size. These *"shorts"* were casually cast back into the ocean again as there was really no reason at all to take the time to measure them. Then there were the *"too-close-to-call"* lobsters that really had to be measured with the gauge. And if any of these lobsters just barely *"missed the gauge,"* when measured from the eye socket to the edge of the carapace, they were measured again from the other eye socket to the edge of the carapace. There were very, very few lobsters among the many measured in this fashion that did actually meet the minimum legal size. These lobsters, of course, were retained as *"keepers,"* all of which is legal since a lobster can be measured from either eye socket to determine legal size.

A lobster being of minimal legal size, however, is not the only test of legality in the Maine lobster fishery. In this fishery, every lobsterman also has to check for three other conditions before he can consider a lobster as being legal to retain. In the case of lobsterman Joe Nickerson, he first inspected the *"tailfan"* to see if a *"V-Notch"* had been previously cut into one of the tailfan's appendages. If there was, then that lobster was returned to the ocean. He also made it a practice of flipping over any larger size lobster to determine if it was a gravid egg-bearing female lobster. If such was the case, or if the lobster was not already V-Notched, he would V-Notch the lobster before returning it to the ocean. Finally, any large size lobsters were measured to determine if they were more than five inches in *"carapace length measure."*

All of the conditions previously described bear out the strongly held convictions of most Maine lobstermen that such retention restraints constitute the present and future brood stock of the Maine lobster fishery, and that any so-affected lobsters should be protected accordingly. In the first situation, the V-Notch was a clear indication that a lobster was at one time, but not necessarily at the present time, an egg-bearing female lobster. In the second consideration, the female had eggs showing, and in the third situation, the gauge revealed that the lobster exceeded the maximum legal size. These are all more or less restrictive conservation and preservation measures that have been adopted over the years by the Maine lobstermen and their respective Lobstermen's Cooperative Associations.

Shorts! Shorts! And More Shorts!

Joe Nickerson's traps were abounding with *"shorts"* despite the fact that the escape vent in each of the two-compartment traps should have allowed for these sub-legal size lobsters to escape from the trap. But it is possible that some of these lobsters had just entered into the *"kitchen"* compartment or into one of the *"parlor"* compartments shortly before the trap was hauled and didn't have ample time to make their escape. Differences in the size and shape of a lobster's carapace, no matter how slight, can also make it difficult for even a sub-legal size lobster to wriggle through the opening in an escape vent. To demonstrate this point, Joe Nickerson seized a sub-legal size lobster and tried to push it through the escape vent opening, but try as he might, that lobster could not be pushed through the opening!

The "Cull Lobster" Population

We have seen that a "cull lobster" can be a lobster with only one large claw or a lobster with no large claws at all. Lobsterman Nickerson's traps contained a large number of the so-called one-claw lobsters and about one or two of the no-claw lobsters, the latter of which are often referred to in the fishery as *"barrels," "bullets,"* and *"paper shells."* While closely observing what was being removed from Nickerson's traps, I became convinced that ***there was a lot of fighting going on down there!*** There were lobsters with regenerating appendages, lobsters with deformities, and lobsters with all sorts of puncture wounds about the claws, the carapace, the walking legs, and the hard shell of the tail section.

There were lobsters with their long branched antennae partially snipped off or severed completely. All of this should probably come as no surprise, because the lobster is known to be a very snappy and pugnacious animal whether it is in the wild on the ocean bottom or while it is being held in captivity. This type of behavior, which can often result in injuries and death, appears to peak during the period of the year when the seawater temperatures are the warmest. They become, at this time, increasingly active and compete very aggressively for the bait in the lobsterman's traps. The American lobster is not only a scavenger, but is truly cannibalistic, and it is so to the extent that it will set upon any weak or vulnerable of its lot, and without any reservation whatsoever in doing so.

Most certainly some of the high cull lobster rate in the population can be attributed to the repeated handling on the part of the lobster fishermen. When hauled to the surface, lobsters come racing up through the water column while imprisoned in their wooden or steel cages - ***and they are fighting mad when they get to the surface, and they don't like it one single bit!*** A previously written scenario seems to be worth repeating at this point: Imagine, if you will, a lobster trap resting undisturbed and in considerable darkness on the bottom of the ocean. Then, all of a sudden - **_Whoosh!_** - and the trap, lobsters and all, gets yanked up off the bottom at a tremendous clip. The trap shoots through the water with a lightning speed at a 45-degree angle, is bounced up on top of the boat's gunwale, and then the lobsters sometimes receive some pretty rough handling by the lobsterman. On the basis of all this, it would appear reasonable to assume that the hauling-up process and the rough handing of some of the lobsters, especially those of sub-legal size, continually hauled to the surface make some sort of contribution to the high incidence of cull lobsters.

Soft-Shell And Hard-Shell Lobsters

Being the month of September, the majority of lobsters pulled to the surface by lobsterman Joe Nickerson were of the hard-shell variety. The hardness and softness of the lobster's shell was easily determined by applying a little lateral pressure to the lower left and right sides of the carapace. When the carapace was depressed easily and was a little *"spongy,"* then that lobster was a soft-shell lobster; if the carapace was not easily depressed and felt quite firm, then that lobster was a hard-shell lobster. Soft-shell lobsters, it may be recalled, are sometimes referred to as *"summer lobsters,"* while hard-shell lobsters are commonly known as *"winter lobsters."*

The Tale of the Lobster

Arnold (Joe) Nickerson is a third generation lobsterman who has always maintained his residency in the Kennebunkport - Cape Porpoise area - and he has always fished out of Cape Porpoise Harbor, a small seaport located about halfway between Portsmouth, New Hampshire and Portland, Maine. Joe, as he prefers to be called, graduated from Kennebunk High School and then served a four year stint in the United States Air Force before settling down to marriage. He is the father of four children who have all gone on to pursue their individual careers. Both his father and grandfather were toilers of the sea and worked many years as Maine lobstermen. Joe Nickerson, like many other Maine lobster hunters, got his start in *"lobstering"* through a father-son relationship. As he modestly puts it, "I liked the ocean as a boy. And, of course, my father *"lobstered"* for awhile, and going with him, I began to like it too."

For the most part, Joe Nickerson's workplace in on the ocean and in his boat, the *"Kori-Anne,"* which he named after his two daughters. He normally fishes about 800 traps as much as possible and especially so when there are enough lobsters to make it worthwhile and when his traps need re-baiting. During the spring, summer, and fall months, he usually hauls about a third of his traps every day, weather-permitting, but during the harshest of the cold winter months he doesn't fish and brings about half of his gear ashore and leaves the remainder to *"set out over the winter."* He only lets those traps to set-over that may have seen their better days, because he doesn't want his top-notch gear damaged or lost as a result of winter storms or commercial dragger fishermen *"raking"* their heavy nets along the ocean's bottom in their pursuit of groundfish.

Like so many other lobstermen still do, Joe Nickerson builds most of his traps during the winter months. In addition to the maintenance he must perform on his boat, his buoys often need touching up with fresh red and yellow paint. He can build a wire trap for about $30.00, but that same trap would cost around $40.00 for some firm to build it for him. The Onset Bay Lobster Company, for example, sells a conventional wire trap for about $40.00, and for another ten dollars they will include the warp, a buoy, and a bait bag.

Bait is a very expensive cost item for the lobster fisherman. Lobsterman Nickerson buys his bait directly from the *"bait man"* conveniently located on the pier at Cape Porpoise, and he goes through about two to three barrels of bait a day "depending on the number of traps hauled and how bad the fish crab's fleas are."

When it comes to conservation measures, lobsterman Joe Nickerson and his fellow lobstermen certainly do not hold back in vocalizing or stating their views on the subject. Some time ago, when he was queried about going from the 3-3/16 inch to a 3½ inch carapace length for lobsters to become of legal size, he responded with an emphatic "No!" He qualified his position by remarking: "I myself wouldn't mind, but do it in 1/64" increments a year. That's about 5% of our catch. Contrary to what the biologists say, 1/16" is over 20% of our catch and we measured our catch. All three of us were 20%. By increasing 1/64" a year, the biologists could watch and see how many more juveniles lobsters were egging out."

[Author's Note: The minimum legal size for lobster retention is now 3¼" carapace length measure]

In addition to the minimum gauge law there are, as we have touched upon, other lobster conservation measures that have been enacted in the hope of protecting Maine's lobster fishery. Joe Nickerson has opinions about them also. The first is whether any lobster, male or female, that measured 5 inches or more in carapace length should be returned to the ocean and not be marketed in the State of Maine. He feels that "doing something is better than doing nothing, and that is what the other states have been doing - nothing!" He goes on to state that "I think the big thing is that it keeps that many more Canadian lobsters out of Maine. The dealers are the ones that want to get rid of the 5" measure. Regarding the matter of *"keeping Canadian lobsters out of Maine,* it appeared that there was a glut of these Canadian lobsters that were shipped into the New England marketplace during the winter of 1999-2000 in anticipation of the Millennium craze, and which had the effect of driving the winter price of lobsters down to record low levels. The time for the Canadian lobster dealers to make their move was perhaps an opportune time, because Canadian fish for lobsters mostly during the cold winter months, while New England and south lobstermen do most of their lobster fishing during the spring, summer, and fall months.

Another lobster conservation measure relates to Maine's *"V-Notch Program"* that is intended to protect the present and future brood stock of the lobster fishery. In this regard, Nickerson remarked: "Yes. And if anyone can't see that, then he's not too smart. You have to give something back. You can't take and take and still have a business. So, if those biologists can't see that, then they aren't too smart."

When queried about the strongest areas of disagreement with some of the research and pronouncements of lobster marine scientists, I struck a raw nerve! "There are many. They don't listen to us for one thing. They think they know it all but they don't. Like the V-Notch program, we know it works. But they've never made a survey to see, and they probably won't.

"Another is this 18 to 20 million pounds of lobster a year we reportedly catch. If you take 20 million pounds and multiply it by $2.50 pound average ($50,000,000) and divide it by 5,000 lobstermen (and there are more than that), then you get $10,000 a year that we make. Hell, my expenses are more than that! The Cape Porpoise, Maine lobster fisherman continues:

"Maine catches probably 75 to 100 millions pounds of lobster a year. Their (the marine biologists) low landing figures are for one reason: to get more grant money from the government to help a supposedly Dying Industry! Or so the government thinks, going by the biologist's figures."

Lobsterman Joe Nickerson goes on to suggest that there are many factors, not just two or three, that affect the supply of lobsters. From the material previously presented in this writing, the reader can deduce that this Maine lobsterman is pretty much on top of his game, and that he has a good grasp of all the factors that have an effect on the American lobster in the State of Maine and elsewhere. For example:

> "Overfishing is probably the big one. You could probably catch as many lobsters with half the traps. You would have a lower mortality rate and less of a cull rate.
> "And, of course, weather plays a big part. The severity of the storms can cause a higher mortality and cull rate.

"Another factor is that very cold winter and spring water temperatures probably only slow down the shedding process. That's why the shedders are later and later each year.

"You also have the draggers that drag lobsters all fall and winter, and you have some lobstermen, and others, selling short lobsters. It all adds up against us in the end. There are probably more short lobsters now than there were 40 years ago, it's just that there are too many hands in the pudding, so they say."

Poaching has been a problem of long-standing and has no doubt persisted since the very inception of the lobster industry. It has been seen as being a major problem along the North Atlantic seaboard of the United States and is said to be even more of a problem in the waters off Labrador, New Brunswick, Prince Edward Island, Cape Breton Island, and Nova Scotia. When asked how he looked upon *"poachers,"* and how Cape Porpoise lobstermen deal with the problem, Joe Nickerson appeared to indicate that the lobstermen themselves take the matter into their own hands:

"Of course no one likes someone else stealing from you, but it happens and they're almost impossible to catch. Some have been caught and usually that person is put out of business or he "loses enough traps" to smarten him up, and he stops it. Wardens have caught some and taken them to court, and the judge slaps them on the wrist and they're back doing it again. So most of the time if a lobsterman catches one of his fellow lobstermen stealing, he takes care of it himself."

The subject of *"outsiders"* and *"part-timers"* cutting into the lobster harvest has been discussed previously and will be touched upon further in Chapter 6 that follows, but it is interesting to take note of how a well-seasoned Maine lobsterman like Joe Nickerson feels about the matter. His candid response reflects, to be sure, the sentiments of most lobstermen everywhere up and down the coast:

> "To be honest with you, it's the people with a good paying shore job, or on a government pension, or a man with plenty of money, that gets the most resentment. They are making good money and we would like to do the same.
>
> "Even though 95% of us would still let you go, every port has its cutters, "men that won't leave you alone." They keep it up for years so it makes it very expensive for that person. Even for a newcomer just trying to make a living, he faces this 5% who are going to harass him and try to drive him out. The rest of us would let him go if he minds his own business."

At the end of a long day of fishing for lobsters with Joe Nickerson, he pulled the *"Kori-Anne"* up to his dealer's *"live car"* in the middle of Cape Porpoise Harbor. He weighed his catch on a large scale, dumped the lobsters into crates, which were then pushed off into the live car. All of this was done without having the presence of the dealer on the scene. When Joe Nickerson was asked about this element of trust, he explained it in this manner:

> "Believe me, it's not all trust. Some dealers have their fishermen put their catch in their own pens, so if there are V-Tails, or illegals, he knows where to look. Some dealers trust everyone, but once in awhile he'll start finding shorts and illegals, and one day he'll crate everyone's catch and check out their lobsters later. If he finds shorts or V-Tails, he'll tell the party to stop and go elsewhere.
>
> "Dealers can and do lose lobsters if they're not careful. Usually your dealer is a good friend of yours and you trust him as such. I can't say this for all areas, but here it seems to be that way."

The Tale of the Lobster

It has been demonstrated that the lobster fisherman incurs considerable expenses in the process of running his business. Included as part of these expenses is that of paying for the services of a *"sternman."* In the case of lobsterman Joe Nickerson, he makes a practice of compensating his sternman on the basis of an hourly wage if he's a youngster working his way through college or merely doing seasonal work. However, if the sternman is working for him full-time throughout the entire lobstering year, he will compensate him on the basis of a percentage of the catch.

Can fishing for lobsters be dangerous? Definitely! There have been accounts of lobstermen falling over board or being pulled overboard by a length of trap warp. There are reports that give credence to the fact that such accidents do happen aboard a lobster boat. And the irony of the matter is that there are many lobstermen who can't even swim! And there have been accounts of lobstermen lost at sea and the reason for their demise has not ever been beyond a reasonable doubt. Take the a situation that occurred in November of 1999 when 49 year old Paul King of Essex, Massachusetts took a small pram out to check on his new 36 foot lobster boat that he had moored at the mouth of the Essex River. He never returned! State police, environmental police, the coast guard, local police, and divers from the local area, and environmental police scoured the area during a four-day search mission.. His body was nowhere to be found. The search was finally called off and the lobsterman was presumed dead. In talking with the Essex (Massachusetts) Chief of Police, it was learned that divers did locate the pram, two oars, and a pair of rubber boots. But off Conomo Point where authorities envisioned that the tragedy took place, the current was running very strong and the water depth was something in the order of fifty feet. At the estimated time of the mishap, the water was extremely cold and the weather was very nasty. It was finally conjectured that the lobsterman's pram tipped over, the lobsterman were overboard with it, and somehow got himself tangled up with some of the debris that littered the bottom. But to the best of the search parties, including helicopters, the body of Paul King was not located as of April 2000. This tragedy serves to emphasize the point that the weather is often the fishermen's worst enemy and that being about and operating a small craft can be extremely dangerous during times when foul weather and a rough are present.

Joe Nickerson related an event that was probably his most harrowing experience during his lifetime career of fishing for lobsters and, as is usually the case in most dilemmas on the ocean, it happened in a split-second:

> "Years back in January, I had a pot warp around my foot and I got dragged to the stern. It was a pair of traps. I had a half hitch around my foot just above the toes, and my foot was up against the inside combing sternrail. Back then we used bottles for toggles and the toggle to that pair of traps was what was around my foot. I had no knife, so I broke the bottle, and with a piece of glass, I cut the rope off. Made me wonder what would have happened if that bottle hadn't been there. Now a knife is on the stern at all times."

Do lobstermen eat lobsters and do they like them? Surprisingly, many of them don't eat lobsters! They probably don't care that much for the taste of the crustacean, or perhaps it's because they don't relish the thought of eating something they have hunted and caught. And how about lobsterman Joe Nickerson? "Yes, I eat

The Tale of the Lobster

lobsters and like them. They are best right out of the pot or in a stew. My family likes them also, but if I eat a dozen a year, that's a lot for me."

[Author's Note: Lobsterman Arnold (Joe) Nickerson III of Kennebunkport, Maine retired from the lobstering business in 1997. He sold his boat to a young lobsterman and all of his trap gear to his son. Now in his mid-sixties, he enjoys a much more relaxed life, but still keeps himself involved in the business by working about 25 hours a week at a local lobster pool in Cape Porpoise, Maine]

Now, we all are aware and hopefully mature enough to realize that any business is made up of people who *"work hard and work smart"*, and there are those people who work on just a *"so-so"* level of performance and do not necessarily apply a great deal of ingenuity to the business. And then there are those people who quite frankly don't put in the hard work, don't always take advantage of the available technology and refinements in the business, and simply *"go with the flow"* of everyday occurrences. This latter group of people is often very complacent in the business environment.

An analogy of sort can be made for the lobster fisherman. There are those in the business that fish more than the average number of traps, work smart, invest in the best of fishing boats to make their work go easier and more efficiently, buy the latest gear and electronic equipment, put in long hours, perhaps go farther out into the offshore lobster fishery, and possess the skill and the know-how in the marketing of their lobsters. They are the *"mainliners"* of the lobster fishery!

There are also the *"middle-of-the-roaders" who* probably fish the average number of traps, do not put in the hard work and the extra-long hours hauling and setting, prefer not to commit to significant investments in gear and associated expenses, and do not take advantage of the cutting edge technology that is available to them.

The third group, the *"bottom-of-the-rungers,"* probably couldn't care less, set out and haul fewer traps, perform less work, work shorter hours, shun serious money investments, and work along at a pace and with the mindset that everything is okay in *"just getting by."*

There is nothing fundamentally wrong with the lobster fishermen that fall into any of the three groupings. It is really a matter of defining one's philosophy of life and to the degree of success one hopes to achieve in the lobster fishing business. This matter is further touched upon in Chapter 6 in the *"Catch" Vs. "Effort"* aspects of the fishery, such as the fishermen, their boats, landings, the number of traps fished, and the like. One final reminder is that the data pertaining to lobster landings, dollar value received, the number of traps fished, and virtually every other aspect of the lobster fishery are reported on the basis of estimates and their resulting averages. There are many lobster fishermen, as is the case for all commercial fishermen, that exceed the average indicators, many who satisfy the averages, and many who fall way short of them. Simply put, the success of any person in business, whatever that business may be, is governed by a personal philosophy involving the work ethic, attitude, self-expectation, risk-taking, and a certain amount of perseverance, self-motivation, and the innate burning desire to succeed in the business.

Now that we've had a little exposure to the Lobster Fisherman, it might be interesting to listen to some *"lobstering talk"* among three Down East Maine lobstermen that is taking place as they stand on the fish pier at Winter Harbor while waiting for the fog to lift. It will be noticed that New Englanders in general don't always bother with the letter "r," such as in the word "short" which comes across as "shot."; they also have a tendency to drop many "g's," as in "lookin'," and are easily prone to shorten some words like "and," which sounds like "n," and so forth. Whatever the reason, whether it be just a passed-on practice or an unnecessary attempt to pronounce every syllable in every word, their language is most often very easy to understand.

"Tom, Dick, and Harry"

"Ya know, Harry, it's a little touchin' on my soul n' patience waitin' for this fog ta lift."
"Well, Dick, ole "Bait Bag Smitty" jus' came back in. He jus' plum gave up, and yelled across ta me, "It's too-thick-a-fog out thar. So I figga it's betta ta wait it out than be sorry."
"Tom, what bait ya fishin' taday?" Tom replies, "Probably the same as you are, Harry, those damned redfish. It's 'bout all that's around right now unless ya wanna go ovah to Bah Haba. They say there's a lotta Atlantic herrin', but it's a hellava drive and I'd havta use my cah 'cause my betta half has the truck. Hell, ya can't haul bait in a cah."
"Dick, how ya catchin' em lately?" "Same as evry spring, Tom, ya know those damned sheddas keep cumin and cumin, but it's money in my pockit, so I keep haulin' n' haulin'." But ya know, there seems to be a lot of those sheddas that don't weigh much more than a pound n' their shells are so soft ya feel like ya pickin'

up a sponge. But ya havta take the bad n' good of it. I've heard that "Dirty Jonesy" is fishin' his geah way out ta get away from the sheddas. Some tell me at Windy's that he is doin' okay and catchin' more hahd shellas n' a good many of em are quite large keepas."

"Dick, "Joe Dipstick," that ovah-ambitious marine patrol offica has bin spendin' a lotta time in the outa haba lately. Have ya heard from the otha lobstamin what he's upta?"

"Nope, but I's bettya he wouldn't know a shot lobsta from a keepa lobsta if one jumped up n' bit im on the knee. But he's gotta right ta be thar 'cause I've heard a lotta talk at the bait shack that somma the fellas have been takin' shot lobstas, n' that fer damned sure is gonna hurt the fish'ry n' all of us in the end. Yup, it could be a hellava problim here in Bah Haba n' along the coas'line."

"Well," Harry pipes up, "that's not the only problem. Ya know, those bug chasi's ar at attit agin. Now those scientists, or whateva ya wanna call em, are lookin' at increasin' the gauge once agin. Hell, they got-it up ta 3¼ inches now! N' this is afta trap limits have been reduced to 600 hundrid traps in most inshore areas.. With the high price of fuel, our profit mahgins are gonna go as low as Davy Jone's Locka!"

Tom patiently waited for Harry to end his complaining about the lobster marine scientists and then remarked, "speakin' a cost n' profit, there were a couple duzin or so people, summah tourists I'd reckin, lined up on the dock then I pulled upta the dock afta fishin' yestaday.

"Hell, they geeked and gawked when I landid the day's catch on the dock. You'd think I had jus' discovered gold or somethin'! It was a nice aftanoon, ya recall, n' they probly thot lobstaing was all fun n' games or somethin'! Ya shud have seen the look on their faces n'talkin' back n' forth with one anotha. Boy, have they gotta lot ta learn, like puttin' upwitha wind n' spray, those wet gloves n' the swell of the ocin, n' all the hahd work that goes inta the haulin' n' settin' the traps. Lobstaing isn't so easy as it seems to be on the surface. We know that, but they don't appeah ta know that."

"Well, guys, the fog's liftin' so I' m gonna head out thar," Tom remarks, and he Dick, and Harry head for their boats, the "Hannah," the "Friendship," and the "Fortitude." As they had already fueled up and loaded bait on board, and as each parted company, Tom said, "Dick, Harry, have a gud day lobstaing. Let's go fishin!"

Tom, Dick, and Harry all returned to Winter Harbor at about the same time, but later than usual because of the morning wait for the fog to lift. They stored their day's catch in their lobster cars in

The Tale of the Lobster

the middle of the harbor, tied their boats up to their moorings, and then got together again for a little chat.

"Tom, howdya hit em taday?" "Dick, I did okay. Ayah. On the orda of 250 pounds or so. Ayah. But "Little Richard" of Notheast Haba pulled up baside me n' said his ketch was way off." He told me, "Tom, I' m of the thinkin' that someones out thar liftin' my geah n' pickin' me clean. Ya know, there's a lotta pleasah boats in the area n' some of em could be haulin' my geah. Then ther's "Lighthandid Louie" workin' the haba n' everybody has their eyes on im. Maybe tamorrah will be a betta day."

Tom says, "Ya know, guys, all my geah wasn't where I set it yestaday either. And my traps didn't ketch as many keepas fer this time of the yeah. I think there's a little poachin' goin' on out thar too, but I think it's more like one of our fella lobstamen, or maybe more than jus' one of em. Maybe we otta go out thar afta dark tanight and see if we can find out what's goin' on. Ayah."

"Yup," Harry says, "n' the otha problem is a growin' numba of em big city people who think they can come up here n' set out a few traps. We gotta discourige that! What ya say we tie thar buoys so they float upside down, n' if they don't get the messig, we'll cut em loose a few atta time to driv'em out? Afta all, thar cuttin' inta our businiz n' it hits us all in the pockitbook in the long run a things."

"Tell ya what, Harry, let's meet heah tanight 'bout 8 n' we'll use my boat ta go out n' see if we can see what's goin' on out thar. Supposed ta be in the 70's n' we can run my boat without any runnin' lights on. If it's "Lighthandid Louie," maybe we'll catchi'im in the act, n' make 'im one misrable SOB." Tom chimes in, "Yup! Gud ideah, n' if it's them pleasah boatas, we'll grab the lobstas, shot or legal, they've poached n' giv'em a stern warnin' neva ta haul lobstamen's geah agin. n' if we find out we're bein' encroachid 'pon by outsidas, maybe we otta tie off a few buoys n' even cutta few traps loose. Too bad we havta take care of these things ourselves, but ya know, those marine patrol officas are spred pretty thin n' they can't be evrywhere all the time." Dick became very disgusted and said, "Yup, yer right, but evin if a patrol offica catches someone in the act n' brings him to court, the judge jus' slaps 'im on the wrist n' let's 'im go. So, what's the point of it?"

The three lobstermen met at the dock at eight o'clock in the evening as planned and began to walk towards Dick's boat. Tom said to Dick, "Ya know, Dick, maybe I'm getting' too old for this kinda stuff." Dick replied, "yeah, kinda gets to yer bones afta a few yeahs, but itsa livin' n' I'm my own boss n' I can run my own businiz as I see fit. That's more than mos' people can say." Harry gestures with his hands and remarks, "Yer right, n' futhamoah we have nothin' else ta turn ta. Lobstaing's all we know so we jus' gotta make the best ofit."

As it so happened, the trip out and back was uneventful. The three lobstermen saw no traces of anyone poaching or lifting their traps. As they tied up Harry's boat on his mooring, Tom says, "Dick, Harry, let's go downta Wendy's n' have a cuppa coffee n' a bagle n' talk 'bout wimin or somethin' otha than lobstas n' lobstaing." Dick says, "I'll vote fer that." Harry says with enthusiasm, "Yeah, let's jus' do it!"

"The Lobster Fisherman"

"Many a summer tourist
has gazed upon the scene,
Of lobsterman and helper
at day's end so serene,
Pulling up at dockside
with lobsters by the score,
Ah, the life of the lobsterman
far, far out from shore.
But it is he the lobster fisherman
whose not the least surprised,
By a trap that's pulled up empty
before his very eyes,
And some of the traps set yesterday
could no longer there be found,
A poacher must have hauled them
and left the fishing grounds.

It is he the lobster fisherman
who has met the test of time,
And has proved himself successful
in his youth and in his prime,
Hail to the lobster fisherman
who braves the weather and the sea,
Who tends his traps and gear
with a sense of humility.

It is he the lobster fisherman
who baits, and sets, and hauls,
Who heads out early every morning
in fog, and rain, and squall,
Then pulling up at dockside
with lobsters by the score,
Ah, the life of the lobsterman
far, far out from shore!"

Robert Delano Martin
Beverly, Massachusetts

Chapter 6. Practices And Conditions That Influence The Lobster Fishery

In this chapter we will delve into a host of subjects dealing with the major practices and conditions that affect and have both a direct and indirect impact on the lobster fishery.

[Author's Note: The topics in this chapter will be basically limited to the lobster fishery in the State of Maine. Certain practices and conditions may very well exist in any or all of the other lobster-producing states, but since Maine is the major lobster-producer in the United States, attention will be focused on the practices and conditions as they apply to the Maine lobster fishery. However, some of the discussion concerning "Catch and Effort" includes data and information relative to American lobster statistics for six of the other eleven lobster-producing states as well as for the State of Maine.]

The items for consideration for inclusion in this chapter cover a wide range of material, and for the purpose of clarity, they are summarized below:

- Lobster Research Program
- Lobster Management Policy Councils
- Lobster Advisory Council
- Lobster Promotion Council
- Lobster Institute – University of Maine
- Lobster Industry Associations
- Closed and Open Fishing Periods
- Intensive and Extensive Fishing Effort
- Minimum Legal Size
- Maximum Legal Size
- Maine's "V-Notch" Program

- "Ghost Traps"
- Poaching
- Escape Vents
- Catch and Seawater Temperatures
- Catch and Effort
- Lobster Migration
- Gear Conflict
- Trap Tags
- Trap and Trawl Limits
- Trap Construction
- Trap Molestation and Removal
- Mortality and Disease
- Licensing
- Law Enforcement.

As a prelude to the topics to be addressed in this chapter, it should be without argument that the survival of the fishery depends to a significant degree upon certain areas of scientific research, findings, and suggestions of the lobster marine scientist. Since this is only one half of the equation, we shall hear about some of the related beliefs and opinions from lobster fishermen and authors whom have written on the subject. As we shall see from the material covered in this chapter, and in Chapter 5, *"The Lobster Fisherman,"* there is often no basis of universal consensus regarding some issues that are considered to be of a somewhat controversial nature.

Lobster marine scientists from many state, national, and provincial jurisdictions have for many years conducted continuous lobster research for the purpose of promoting the viability and preservation of the American lobster fishery. The State of Maine has always been pretty much the standard bearer in this research and continues to be so up to the current time. The Maine Department of Marine Resources (DMR) has for many years had the talent and dedication of now-retired James C. Thomas, and for many years since has had his work continued by Jay S. Krouse who was recently retired after many years of service with the Maine DMR. Their predecessors and others that followed Thomas and Krouse have been unrelentless in their research of the American lobster.

The lobster marine scientists, often referred to as marine biologists, from time to time have conducted lobster research in such areas as catch vs. effort, molting, migration, cull rates, escape vents, mortality rates, reproduction, seawater temperatures, and a host of other topics related to the lobster. Probably one of the most important goals of the lobster marine scientist is that of making recommendations to the legislatures, the lobster councils, and the lobster fishermen with regard to any practices or conditions that will hopefully ensure that the American lobster will continue to thrive and not become an *"endangered specie."*

It might be said, however, that despite the good intentions of the lobster marine scientists, they and their research data are often ignored and criticized by many lobster fishermen. The act of friendly persuasion goes just so far and does not always set too well with many a lobsterman. There have been, in fact, occasions when some of their lot have become somewhat hostile and argumentative when alluding to the research, findings, and recommendations put forth by these researchers of the American lobster. Some of their lot have gone so far as referring to lobster marine scientists as *"bug chasers."*

In an interview with James C. Thomas, he contended that the fisherman's approach is to make money, and he doesn't want anything to happen to jeopardize what he's earning now. He's not too anxious to lose something now to gain something later." Thomas, a very likable person, was very quick to add, "but I understand where they're coming from." He claims that he and other marine scientists operate from the viewpoint that they only have a voice and that their role is to serve the fishermen in an advisory capacity. He further takes the position that "it is the fishermen and the legislators that control the lobster fishery and the regulations of the fishery." In the Maine DMR Lobster Informational Leaflet #3, the role of the marine scientist is stated with a little different twist: "We know that our opinions are not popular with some of the people in the lobster fishery. Nevertheless, we must present the facts as we see them. Our role is only to advise and then the fishermen through their legislators actually decide the course of action." It would seem reasonable, therefore, that any controversial issue that comes along should be considered with a positive attitude by the lobster fishermen and the state agencies that regulate them.

[Author's Note: The following discourse relating to Maine's Lobster Research Program, Lobster Councils, Lobster Institute, and Lobster Industry Associations has been included to illustrate the depth of their purpose and goals. Their functions, together with the functions of the lobstermen themselves, have one sole purpose in mind - the preservation of the Maine lobster fishery. Massachusetts and other lobster-producing states have some similar organizations in place, but certainly not of the magnitude as those of the State of Maine.]

Insofar as the Maine Department of Marine Resources publicly states, the following language concerning the purpose and goals of the Maine DMR are set forth in their publication, *"Compilation of Lobster Laws and Related Regulations:"*

Lobster Research Program

"The commissioner shall establish a program of lobster research within the Bureau of Marine Sciences. The purpose of this program will be to develop reliable scientific information for use in management decisions."

1. Research. The lobster research program shall include investigation of lobster population dynamics, reproductive behavior and migration patterns. Specific attention shall be given to evaluating the impacts of the State's V-Notch program on the reproductive potential of lobster stocks.
2. Policy investigations. The commissioner shall develop in the lobster research program the capacity to systematically analyze the effects of conservation and management options. The analysis includes both the biological and economic components of the fishery. Options for policy analysis include, without limitation, changes in the lobster measures, seasons, limitations on effort and limitations on entry to the fishery. Analysis of these options shall be conducted cooperatively with the industry and the Lobster Advisory Council.
3. Data collection. The commissioner shall continue the lobster fisheries data collection undertaken by the department since 1967. Continuity of data collection shall be ensured.
4. Cooperation. The commissioner shall cooperatively develop and coordinate the lobster research program with the University of Maine and the lobster industry.
5. Report. The commissioner shall prepare an annual report to the legislature setting out the accomplishments of the previous year and an updated, 5-year research plan for future activities with proposed budget requirements. The report shall be reviewed by the Lobster Advisory Council prior to submission to the Legislature. The report shall be submitted to the joint standing committee of the Legislature having jurisdiction over marine resources on or before March 15th of each year.

6. Funds. All federal and state funds obtained and used by the department for lobster research shall be utilized to achieve the objectives of this subchapter."

Lobster Management Policy Councils

The lobster fishing industry in Maine is extensive in terms of ocean acreage and miles of shoreline. The thousands of lobster fishermen and the tens of thousands of lobster traps have over the years called for the creation of certain establishments to aid in the management and control of this burgeoning industry. One of these establishments is the Lobster Management Policy Council, which is essentially made up of seven individual councils, one for each of the seven "zones" along the Maine coastline. Each of the separate councils is comprised of elected licensed lobster fishermen and appointed state legislators. The key role of each of the lobster management councils is to provide recommendations as to the overall management of the fishery and with emphasis on three specific areas of concern: 1) the number of traps allowed to be fished in each zone and the time allotted to comply with those trap limits, 2) the number of individual traps that can be fished on a trawl, and 3) the time periods that are open and closed for the fishing of lobsters. It is therefore obvious that any recommendations that are forwarded for consideration by the Commissioner of the Maine DMR have provided the thousands of Maine lobstermen with an input and a voice in the way the lobster fishery is to be regulated in the State of Maine.

Lobster Advisory Council

The composition of the Lobster Advisory Council is more diverse and perhaps more influential than that of the Lobster Management Policy Councils. The Lobster Advisory Council is represented by more segments of the lobster industry and is unique as to the makeup of its members. The Maine DMR in its publication, *"Compilation of Lobster Laws and Related Regulations,"* sets forth the following appointments and composition of the Lobster Advisory Council: "The Lobster Advisory Council…known as the "council," consist of the following members:

    A. One person from each lobster management council… Each lobster management policy council shall choose by majority vote a member to serve on the council.
    B. Two persons who hold wholesale seafood licenses and are primary dealers in lobsters, appointed by the commissioner.
    C. One person who is a member of the general public and does not hold any license under this subchapter, appointed by the commissioner; and
    D. Three persons who hold lobster and crab fishing licenses and who are not members of lobster management policy councils…appointed by the commissioner. Each person appointed…must reside in a different county."

The purpose of the Lobster Advisory Council, as stated in the above-mentioned publication, is as follows:

"The lobster fishing industry is one of the most important industries in the State because of its contribution to the economy and also because of its unique social, historic and cultural contributions to this State's quality of life.

This subchapter is enacted to help conserve and promote the prosperity and welfare of the State and its citizens and the lobster fishing that helps to support them. This subchapter will accomplish these goals by fostering and promoting better methods of conserving, utilizing, processing, marketing and studying the lobster."

Lobster Promotion Council

As far as the Maine lobster industry is concerned, we have discussed seven Lobster Management Policy Councils and one Lobster Advisory Council. But that is not the end of the existence of "lobster councils" in the State of Maine. Not mentioned thus far are the activities of a third type of "lobster council" in the State of Maine, the Lobster Promotion Council, which appears to have as its main objective the role of promoting and marketing the Maine lobster within the state, within the United States, and throughout the world. This, of course, is a vital undertaking for the state's economic well being and to the thousands of Maine lobstermen who labor and often struggle under difficult circumstances to support themselves and their families through the harvesting of lobsters.

In its publication, *"Compilation of Lobster Laws and related Regulation,"* the purpose of the Lobster Promotion Council is stated as follows:

"...the "council" is created to promote and market actively Maine lobsters in state, regional, national and international markets. The council shall draw upon the expertise of the Maine lobster industry and established marketing firms to identify market areas that will provide the greatest return on the investments made by license holders and undertake those media or promotional efforts that represent the most cost-effective use of a limited promotional budget. The council shall remain responsive to the Maine lobster industry, conduct its business in a public manner and undertake marketing efforts that promote the quality and full utilization of the product and the unique character of the coastal Maine lobster fishery."

The makeup of the Lobster Promotion Council organization is very diverse, at the heart of which is the Commissioner's appointment of three council members, with limited tenure, from each of three established districts in the state. These council members are appointed from a list of nominees prepared by the Lobster Advisory Council.

"The council consists of 9 voting members appointed as follows:

A. From the western district of the State, consisting of lands located between the Piscataqua River and the Kennebec River, 3 members...
B. From the mid-coast district of the State, consisting of lands located between the Kennebec River and the Penobscot River, 3 members...
C. From the eastern district of the State, consisting of lands located between the Penobscot River and the St. Croix River, 3 members..."

These voting members, as appointed by the Commissioner of the Maine DMR, must satisfy certain qualifications in order to be eligible for appointment.

"Qualifications of members: From each district, 3 members must be appointed who meet the following criteria:

A. One person who is a full-time harvester and who has held a valid lobster and crab fishing license for at least 5 consecutive years;
B. One person who is a dealer or pound operator and who:
   1. Has held a valid wholesale shellfish license or transport license for a period of at least 5 years; or
   2. Is the manager of, or an officer in, a business entity operating in the State that holds a valid wholesale shellfish license or lobster transport license; and
C. One person who is a public member."

A person is eligible for appointment to the council from a district only if that person is a resident of the district or if that person's place of business is located within the district.

From the detailed list of powers and duties to be discharged to the council, I have selected the following four, for the purpose of brevity, as being of the most importance to the success of the Maine lobster fishery:

"The council may:

    A. Undertake promotional marketing programs in cooperation with the lobster industry;
    B. Promote national and international markets for lobsters harvested or processed in the State;
    C. Provide material and technical assistance to persons seeking to market lobsters harvested or processed in the state;
    D. Conduct other efforts as determined necessary to increase the sale of lobsters harvested or processed in the State."

The council has the responsibility of hiring a full-time executive director and staff employees necessary to carry out its responsibilities. However, these employees "serve at the pleasure of the council," are not employees of the state per se, and receive salaries and benefits determined by the council. The Lobster Promotion Council is funded from revenues received as "Promotion Surcharges" levied all the various types of licenses issued by the Maine Department of Marine Resources.

Lobster Institute – University of Maine

There is only one state from among all of the lobster-producing states that provides a "cooperative program of research and education" concerning the lobster fishery. It is known and recognized as the Lobster Institute – University of Maine. The Institute has prepared a very interesting and informative publication, which has been universally made available through access to its Web Page over the Internet.

The role and purpose of the Lobster Institute at the University of Maine – Orono are clearly defined in this publication, and are provided here in its entirety:

"The LOBSTER INSTITUTE has evolved over the years to become the major formal mechanism of communication among harvesters, dealers, pound owners, and processors in North America; and between industry and scientists/resource managers throughout the world.

"The LOBSTER INSTITUTE is a cooperative program of research and education with the lobster industry at the University of Maine. Information generated through the Institute about the American lobster (Homarus americanus) is intended to help conserve and enhance the resource, thereby ensuring the continuance of a strong and healthy industry in the state and the region.

"As part of the University of Maine's Research and Public Service efforts, the Lobster Institute is located on the Orono campus. This location is central to the resource as well as the region served by the Institute, which includes the northeastern coastal states and the Maritime Provinces of Canada. The University provides an open, objective setting for the free exchange of ideas by all those interested in the lobster resource and industry.

"The LOBSTER INSTITUTE identifies practical problems of concern to the industry and seeks solutions to these problems. Some solutions may be found through quick-response projects, while others require long-term research programs. By identifying the research priorities of the industry and providing industry assistance to researchers, the Institute links industry expertise with academic resources to promote a better understanding of the lobster and our impact on it.

"Some research priorities of the Institute are projects on:

- Lobster ecology, biology, behavior, and population dynamics
- Larval/juvenile lobster densities, recruitment, and habitat
- Health and disease
- Hatch and release/enhancement
- Marketing and economics
- New product development.

"Other research projects have included: V-Notch surveys, claw band testing, ghost traps fishing impacts, lobster/fish farm interactions, genetics, industry opinion survey, taste tests on meat, and artificial bait.

"The information generated through Institute-sponsored research programs is communicated freely in a variety of ways including outreach education conducted by faculty, students and industry members, as well as conferences, seminars, and workshops.

"The Institute is primarily funded by contributions from the industry itself, and through private donations by friends of the industry. Other funds and services are provided by the University of Maine, research granting agencies and institutions, and private foundations.

"The Lobster Institute offers a wide range of educational programs. These include seminar programs and workshops for industry members and scientists throughout the region. The Institute also has an extensive lobster fishery library with nearly 400 journal articles, research reports, and informational pamphlets about lobsters.

"An industry exchange scholarship program has been established through the Institute which provides opportunities for industry members to visit other regions, states, provinces and countries to encourage communication, understanding, and cooperation among industry factions.

"The Lobster Institute seeks to serve the region's lobster industry, and to provide its friends in the general public, who are interested in the industry and its future, with programs and information that are both timely and meaningful. If you are interested in learning more about the Lobster Institute and how you can support its goals and programs, or if you are seeking information about particular areas of interest, please feel free to contact us with the information provided on our address page."

Lobster Industry Associations

The Maine Lobstermen's Association (MLA) was formed in 1954 and states the claim of being the oldest and largest fisheries association in New England. With its headquarters at Damariscotta, the MLA has approximately 1200 members with an outreach that extends from Kittery at the New Hampshire/Maine border to Lubec at the Maine/Canadian border. The association is more territorial than the other two lobstermen's associations in Maine in the sense that it serves and represents lobster fishermen located from north to south along the Maine coastline.

The stated purpose and goals of the Maine Lobstermen's Association, as excerpted from a publication provided by the Maine Lobster Promotion Council, are as follows:

- Gives Maine's lobstermen a voice and influence at the highest levels of government
- Is dedicated to the conservation and management of the lobster resource
- Encourages community-based management efforts
- Has worked hard to pass legislation such as V-Notch protection, maximum gauge measure, and landing by conventional traps only
- Supports reasonable sea sampling and fosters sound scientific research
- Works to restore and maintain marine resources
- Communicates with industry members through a monthly newsletter
- Offers a variety of insurance options to its members.

The Downeast Lobstermen's Association (DELA) has its headquarters in Deer Isle, the DELA and has approximately 700 members in an area that is referred to as "Downeast Maine" that extends, in accordance with whomever one talks with, from Portland to the Canadian border.

The stated purpose and goals of the Downeast Lobstermen's Association, as excerpted from a publication provided by the Maine Lobster Promotion Council, are as follows:

- Maintain a sustainable resource, utilizing sound scientific data, as well as common sense in developing a long term management plan for the future of the industry
- Ensure survival and participation of the small fishing villages along the coast of Maine
- Maintain the fishery as a fleet of owner/operator boats
- Bring common sense into the political arena
- Educate about the conditions that have created today's lobster industry.

The Southern Maine Lobstermen's Association (SMLA) was formed in 1973. The SMLA has its headquarters at Kittery and has an outreach along the coast that extends from Kittery to Portland. The SMLA has only 100 members at the turn of the 20th Century, but its president anticipates accelerated membership growth in the future. The stated purpose and goals of the Southern Maine Lobstermen's Association are less in scope, though undoubtedly no less important, than those of the other two lobstermen's associations in Maine, the Maine Lobstermen's Association and the Downeast Maine Lobstermen's Association. The stated purpose and goals of the Southern Maine Lobstermen's Association, as excerpted from a publication provided by the Maine Lobster Promotion Council, are as follows:

- Provide a voice to the lobster harvesters that fish from Kittery to Portland
- Represent the interests of the harvesters from southern Maine on management and regulatory issues
- Provide members with a "store" where they can purchase rope, buoys, and other supplies at discounted rates.

Up to this point we have discussed the asserted purpose and goals of the three lobstermen's associations in the State of Maine. That, however, is not the extent of the lobster industry associations in the state. There are three other lobster-related associations in operation that function on a more commercial aspect of the fishery.

The stated purpose and goals of these associations, as excerpted from a publication provided by the Maine Lobster Promotion Council, are as follows:

Maine Import-Export Lobster Dealer's Association (MIELDA)

- Deals with issues pertinent to Maine lobster dealers whether they are strictly domestic or global in nature
- Represents dealers on state and national legislative matters
- Fosters research about Homarus americanus
- Serves as a liaison between the association and other industry groups
- Sits on several of the Maine Lobster Promotion Council's committees.

Maine Lobster Poundowner's Association (MPLA)

- Dedicated to the storage of lobsters during the fall months, and sells then to the market from January through April
- Proper feeding, handling and storage in pounds
- Is an advocate and sponsor of lobster research.

[Author's Note: Maine is reported to have approximately 65 active lobster pounds throughout the state. "Lobster pounds are enclosed tidal areas or coves where substantial quantities of live lobsters can be stored and kept healthy for long periods of time." The lobster pounds in Maine are reported to have a capacity to hold five million pounds of lobsters!]

Maine Lobster Processors, Inc. (MLPI)

- Dedicated to delivering value-added Maine lobster products of the finest quality possible
- Established to assure the uniform high quality of frozen, value-added lobster products
- Produced under strict quality standards
- Quality assurance, furthering technological advances and new product development are critical to growth in this sector of the industry.

Closed And Open Fishing Periods

There are two significant *"closed"* lobster fishing periods throughout Maine coastal waters. The first closed period occurs during the summer months. From June

1st to October 31st, both days inclusive, lobster traps cannot be raised, hauled, or transferred during the time span of ½ hour after sunset to ½ hour before sunrise.

The second closed lobster fishing period occurs during weekends. During the period of June 1st to August 31st, both days inclusive, lobster traps cannot be raised, hauled, or transferred from 4 PM on Saturday afternoon until ½ hour before sunrise on the following Monday morning. This prohibition, therefore, means that the fishing for lobsters on Sundays in Maine coastal waters is prohibited during this three-month period in the summer.

The State of Maine has also established certain areas of its coastal waters as *"Lobster Conservation Areas."* Most notable among these is the Monhegan Lobster Conservation Area where there is both a closed and an open period for lobster fishing. The *"Closed Period"* extends from June 26th to November 30th on an annual basis. The *"Open Period"* extends from December 1st to the following June 25th. This open period consists of 180 consecutive days, as established by the Maine Department of Marine Resources, and based upon the preference of 2/3 of the individuals registered to obtain Monhegan Lobster Conservation Area *"Trap Tags."* Monhegan Island is a small piece of land that is approximately 17 miles east of Boothbay Harbor and 20 miles south of Port Clyde. Summer ferry service is available round-trip from Boothbay Harbor, and passenger service is available round trip from Port Clyde.

There is also a body of water adjacent to Criehaven Island that is closed and opened at the will of the lobster fishermen on the island. The area is closed or opened whenever a majority of the lobster fishermen so petition the Commissioner of the Maine Department of Marine Resources.

Of final consideration is a small portion of the York River, from its mouth to a certain point on the river where it meets with the ocean. Lobster fishing is prohibited in this area.

Intensive And Extensive Fishing Effort

The thorough research that has been undertaken by lobster marine scientists and university people working with lobsters has, at times, given rise to the very possible concern that the American lobster fishery in the United States and Canada might be headed for a steep decline. While the Preface and Chapter 1 of this book has delved into this subject matter to some degree, it is important to take note that in the late 1970's some observers claimed that they were looking at all of the classic signs of a downfall of an industry; there was going to be a crash because of the fishery becoming depleted. Similar forecasts and predictions have been pronounced as far back as the earliest years of the 20th Century, and have been continually reiterated ever since. In 1909, Francis H. Herrick addressed the issue head-on:

> "The causes of the decline of the fishery are plainly evident. More lobsters have been taken from the sea that nature has been able to replace by the slow process of reproduction and growth. In other words, man has continually gathered in the wild crop, but has bestowed no effective care upon the seed. Up to the present time every effort to check the constant and ever-increasing drain upon this fishery has signally failed, which shows that either the laws are defective or that the means of enforcing them are insufficient."

Kendall Merriam in his book, *The Illustrated Dictionary Of Lobstering,"* had some interesting viewpoints about the Maine lobster fishery and the Maine lobstermen:

> "The steady increase in demand has been met as best it can by the fishermen who put out more and more traps, built better boats to fish in harder weather, and hire help to make the work go more efficiently. All this, of course, has put an incredible strain on the lobster supply. Some men have teams fishing 3,000 traps. Some areas are almost completely fished out."

Twenty years or so ago, in Yankee Magazine, writer Lew Dietz had this to say about *"the typical"* Maine lobsterman, his prey, and his harvest:

> "He reaps but neither sows or fertilizes. Only when faced with steadily decreasing stocks does he consider the very possibility of a depleted resource. Even then, the typical Maine lobsterman is reluctant to compromise his independence by banding with his fellows to do something about it."

Up to this point we have been jockeying around and about the premise that more and more traps are being set out every year and that this intensive fishing effort is trapping about 90 to 95 percent of the total legal size lobster population every year. While *"over-fishing"* has been regarded as a detriment to the survival of the industry, certainly one of the foremost concerns of the lobster marine scientist, and others, is associated with the *minimum legal size* of the American lobster.

Minimum Legal Size

Ever since the American lobster was discovered to a be a very sought-after and marketable commodity, the crustacean has been measured in many different ways to determine if it is of legal size to be retained by the lobster fisherman and sold as such. The question of *"how long should a lobster be?"* in order for it to be considered legal size has probably stirred up more debate and controversy than any other topic relating to the management of the fishery.

In 1874, the State of Maine enacted the first law of its kind to regulate the legal size of the American lobster. In times past, the lobster was measured in many different ways, such as from the distance between the outstretched claws to the outstretched tail. But in 1919, Maine decided to go another route and started to use the *"carapace length"* (CL) as the standard unit of measurement. That criteria has found acceptance up to the present time. The *minimum* legal size length was then set at 3½ inches CL measure, as measured from the rear of the eye socket to the <u>nearest point</u> of the posterior end of the carapace shell. Two years later, in 1921, the 3½ inch carapace length measure remained the same, but the type of measurement changed and was measured with a gauge from the rear of the eye socket to the rear end of the body shell <u>and along a line parallel to the centerline of the body shell</u> (the carapace). [This centerline, biologically called the *median furrow line* can be observed through visual inspection of the lobster's carapace.] Then in the early 1930s, the minimum legal size was reduced to 3-3/16 inches CL measure, only to be increased once again to a CL length measure of 3-1/8 inches in the mid-1940's. In 1958, the minimum size was restored at 3-3/16 inch CL measure and remained that way for about thirty years. Following prolonged and often heated debates and deliberations, it was not until 1988 that the law was changed to increase the minimum legal size to 3¼ inch CL measure, and that is the legal minimum size at the present time. Maine was going after an even larger increase in the gauge, 3-5/16 inch CL measure, but industry resistance was so forceful that this objective was never reached. So, to sum matters up, the current minimum legal size for all lobster-producing states is now 3¼ CL measure.

John T. Hughes, long-time former director of the Massachusetts State Lobster Hatchery and Research Station, had this to say about the minimum legal size of lobsters: "Almost 90% of the lobsters landed are just legal size, and less than 10% of these have reached maturity. What kind of nuts are we?" In his statement, Hughes was referring to the then minimum legal size of 3-3/16 inch carapace length measure. Traveling to "Downeast" Maine, Kendall Merriam shared a similar point of view: "though there is a disagreement at what age a lobster breeds, it seems fairly clear that the lobsters of seven years, or when they may be caught, have probably bred only once if at all. This is confirmed by the statement of one lobsterman that they are catching ninety-five percent of the legal lobsters in any given year. The lobsters are being fished out, possibly to extinction, unless some measures… such as an increase in the legal size, are enacted."

Maximum Legal Size

Maine was the first lobster-producing state to incorporate both a minimum and maximum legal size limit. This is often referred to as the *"double gauge law."* The maximum size limit is set at more than 5 inches, CL measure, and as such any lobster, male or female, is required by law to be returned to the ocean. The maximum gauge law, as we have seen, is predicated on the widely held belief that these lobsters are good brood stock and that the females are good egg-producers in terms of their *"fecundity"* (egg-bearing capacity). The inference here is that the larger the female lobster, the greater the quantity of eggs will be released into the *water column* at some future point in time. Francis H. Herrick addressed the matter in 1909 when he had published the *"Natural History of the American Lobster."* In that publication, he went to great length to expound, among other conservation matters, that the minimum gauge would protect sexually immature lobsters, while the maximum gauge measure would lend itself to the protection of the large egg producers. An abundant supply of eggs was probably the second most serious of Herrick's concerns about the lobster fishery. He argued that, "by legalizing the capture of the large adult animals, above ten inches in length, we have destroyed the chief egg producers, upon which the race in this animal, as in every other, must depend." So it would seem that the message is quite clear: protect the smaller lobsters to at least the time that they have reproduced <u>*at least once*</u> and protect the larger lobsters so that they might continue to reproduce and enhance the lobster fisheries in the months and years to come.

Maine's maximum gauge is probably the reason why the Maine lobsterman doesn't get all that excited or feels *"a rush"* when he hauls a trap aboard that contains a large *"jumbo size"* lobster! He has a suspicion, even before removing that lobster from the trap, that it exceeds the maximum gauge (or that it is a *"berried female lobster*, or a female with a *"V-Notch"* cut into one of the blades of its tailfan. But the lobsterman will not begrudge giving that lobster back, because he has that deep-down personal conviction that he's doing his part in protecting the future brood stock of the Maine lobster fishery.

Maine's "V-Notch" Program

A somewhat controversial lobster preservation practice has evolved over the years in the State of Maine. This practice, as we have touched upon previously, involves the *"V-Notching"* of egg-bearing female lobsters, a procedure that Maine lobstermen themselves introduced into the fishery. It was another in their line of contributions aimed at preserving the future lobster brood stock in Maine waters. But some lobster marine scientists and lobstermen were not always in total agreement about the value of V-Notching lobsters. They concluded that the disadvantages outweighed any advantages gained and that it was a doubtful preservation measure at best. This attitude on the part of the researchers brought about the anger and resentment of many of Maine's lobstermen. In 1984, Edward A. Blackmore, then a working lobsterman and president of the Maine Lobstermen's Association, had some curt words to say regarding the opposition to the V-Notch by lobster marine scientists and even by some working lobstermen:

> "I have bad news on the V-Notching regulation that I have been trying for almost three years to get accepted as part of the Federal Lobster Plan. The V-Notch proposal was thrown out by the National Marine Fishery Service and rejected by the Secretary of Commerce in his review.

"There were five New England Council members who signed a minority ought not to pass report and sent it to the Secretary of Commerce. I feel that this was an irregular procedure and certainly not sanctioned by the New England Council. Be that as it may, the V-Notch proposal is out of the Federal Plan. The bureaucrats have won another round...

"For your information, the gang of five who signed the minority report were Ted Spurr, New Hampshire; Herb Drake, New Hampshire; William Lund, Connecticut; James Costakes, Massachusetts; Les Smith, Massachusetts. Ted Spurr has led this opposition from the beginning...

"We have really had it when a state like New Hampshire can call the tune to lobster management. We have more lobstermen in one harbor in Maine than there is in the whole state of New Hampshire. The New England Fishery Management Council is proving to be a poor tool for fishery management for Maine's fishermen."

It was not until 1948 that *"berried"* female lobsters were targeted and *"marked"* by cutting a V-Notch into one of the appendages that make up the *tailfan* of the lobster. The latest V-Notch law in Maine was enacted in 1992: "V-Notch means a straight sided triangular cut tapering off to a sharp point with a depth of at least 1/8." While earlier V-Notching procedures were <u>directed at the state</u> buying *"seed"* female lobsters, and returning them to the ocean after they were "notched," the present day practice is such that the lobstermen are the ones who do the notching, and they do this aboard their boats and immediately return the V-Notched lobsters to the ocean.

<u>*"Ghost Traps"*</u>

Severe turbulence on the bottom of the ocean can cause trap gear to become separated from the buoy, or buoys, on the surface. Specifically, this can occur if the warp becomes separated from its buoy, if the warp becomes separated from the trap's bridle warp, of if the bridle warp becomes separated from the trap itself. Any one of these three happenings means that the trap sitting on the bottom *cannot* be hauled to the surface. The trap, therefore, becomes what is known as a *"ghost trap."* If a single trap is being fished, rather than a string or trawl of traps, the loss to the fisherman would be that one trap and most likely the buoy, the toggle buoy, and the warp that connected all the gear together. In most cases, the trap warp simply descends into the water column and the main buoy and the toggle buoy (if one is being used) float off and bob about on the surface where they are usually promptly picked up by pleasure boaters who like them for "decorating" their back yards and fences.

Should a lobsterman be fishing a *"string"* or *"trawl"* of traps, there would be a buoy at each end, thereby reducing the risk of losing all of the trap gear should one of the buoys be separated from its associated trap warp. Thus, should one of the buoys be separated from the warp at one end of the string, the buoy at the other end of the string would still be intact and the lobsterman can haul from that end and salvage his traps. If a buoy is cut off from the warp leading to the trap at each end of the string, that would amount to double jeopardy because that lobsterman will have lost the complete string of traps which could amount to as many as eight to ten traps and sometimes even more.

Occasionally, wire traps get tossed and tumbled about on the bottom of the ocean and then are often *"washed ashore"* and onto rocky outcroppings and sandy beaches. However, most traps that become separated from their buoys will remain on the bottom as *"ghost traps"* and will most likely contain lobsters and other species of marine life. The ghost trap will also attract lobsters entering such a trap for the first time, probably seeking only a place to hide. If it were not for the incorporation of *"escape vents"* being used in wire traps, sub-legal size lobsters could not escape the trap and would in some instances be cannibalized and die. In any event, all lobsters caught and forever remaining in ghost traps are a significant loss to the lobster fishery. It should be pointed out, however, that all lobsters do not meet their demise by being confined to ghost traps. The introduction of the *"biodegradable panel"* was a very important scientific invention targeted to ameliorate the problem of lobster entrapment. The panel is made of vinyl material that is approximately four square inches and is located at the top of the *"parlor"* compartment. The panel is hinged onto the trap by wire loops that over time rust out, and thereby allowing many of the entrapped lobsters to escape.

It has been estimated that about 20 to 25 percent of all traps annually fished in Maine waters are lost at sea and become ghost traps. If approximately two million traps were being fished in Maine waters in 1992 (a very conservative estimate), and if 22½% of these traps ended up as ghost traps, and each trap and its associated gear costs about $45, then this annual loss of trap gear would have cost the Maine lobstermen more than 20 Million dollars!

The huge amount of trap loss does not always occur as a result of severe weather conditions. There are careless boaters around and about who pay little attention to buoys bobbing about on the surface. The propeller of any type of watercraft is very adept at quickly severing the warp that connects the buoy on the surface to the trap on the bottom. A most common occurrence, however, is when a trap warp becomes entangled around the propeller or the propeller shaft. The lobsterman might stand a 50-50 chance of not losing a trap if a boater shows some sort of consideration and takes the time to untangle the warp with the buoy still attached. Even if he has to cut the warp, and then re-ties it, then the buoy and warp can be tossed back into the water again and that trap gear would not be lost and become a part of the ghost trap population. But there are a certain uncaring and impatient boaters on the waterways who would simply cut and release the warp, and that trap becomes a ghost trap forever!

Some lobstermen, it must be admitted, only add to the ghost trap problem by setting their gear, usually *"singles,"* in harbor channels. In addition to the loss of trap gear, these buoys can present a clear and present danger to all boaters who navigate in deep water channels. The removal of traps in a channel can be ordered by the Army Corps of Engineers which is responsible for keeping federal channels clear for navigational purposes. A recent incident occurred when the Engineers found the channel at Cape Porpoise Harbor strewn with lobster buoys. After several discussions with the lobstermen fishing out of the harbor, the traps were relocated and the Engineers, compromising with the fishermen, reduced the width of the channel from 100 feet to 50 feet.

The loss of lobsters stranded in ghost traps is significant. A two-year study, from 1971 to 1973, was conducted by the Maine DMR in an attempt to determine the impact of ghost traps stranded on the bottom off Jonesport, Maine. The Jonesport site was selected because the bottom was muddy and featured very little rocky substrate that lobsters usually prefer for cover and protection. Lobstermen considered this as <u>not</u> a good habitat for lobsters. But there were lobsters there. The Jonesport Study employed the use of tagged sub-legal size lobsters that were set in wood traps with a 33mm spacing between the lathing. The traps were periodically brought to the surface and provided many revelations with reference to the happenings that took place during this study period. Occurrences such as *"molting," "cannibalism," "mortality" and "recruitment"* were observed and documented. The following summary highlights what happened to these lobsters over the two-year study period:

First Summer-Fall Season	Second Summer-Fall Season
43% could not be accounted for	22% could not be accounted for
25% remained captive	55% remained captive
20% escaped and were later recaptured	5% escaped and were later recaptured
12% were cannibalized	18% were cannibalized

One of the trapped lobsters in the Jonesport Study had remained in the same trap for 22 months, during which time it molted at least once and could be considered as a *"recruit,"* a term that the lobster marine scientist uses to denote a sub-legal size lobster that has just recently molted and has grown enough in size to be then measured as a legal size lobster. As can be observed from the statistics presented above, there is a high rate of cannibalism when lobsters are imprisoned in ghost traps, and that these lobsters are especially vulnerable to such cannibalism just after going through a molt and during the subsequent period of time that will be required for them to acquire new hard and thick shells.

Robert Delano Martin

Poaching

The generic definition of *"poaching"* is *"to take game or fish illegally."* While the lobster is not a "fish," it might be called "game" since it is an animal. The crying shame about poaching is that the one who steals from a lobsterman's trap, and especially the pleasure-boater, cares little about the size of the lobsters he is stealing. *Shorts, legals, jumbos, V-Notchers,* and *eggers* are all fair game – and that person doesn't care, and most likely just doesn't realize how important these lobsters are to the fishery, let alone to the financial well being of the lobsterman who depends on them to make a living. It is a continuing practice that has gone on for years and years and which begs for some kind of resolution. But owing to the limited deployment of law enforcement personnel, a solution to this problem does not appear to be visible on the horizon. There's a big ocean out there and there are simply not enough enforcement officers along the shore to effectively guard against the nasty practice of poaching. And even if a person was caught stealing from the traps of others, and was arrested and brought before the court for such an offense, the judge would in all likelihood just give him a small fine, or no fine at all, and send him off with a warning.

Ken Martin, a former lobsterman who I have been fishing with out of Beverly Harbor, had this to say about poachers: "The estimates of lobsters taken this way are understated – plus the constant *"jigging"* of traps *"by pleasure-boaters,"* etc., adds to the problem." Is this practice of poaching a serious problem? Yes! The *"jigging of traps,"* according to Ken Martin, has been a common practice over the years by unscrupulous pleasure-boaters. He claims that "some of these people are known locally on a first name basis," and often being modified in behavior by the effects of alcohol, they think it is fun to go out and "haul-a-few." He adds that "the activity is <u>tremendous</u> during summer. I have heard men in bars bragging about the "bushels" of lobsters they and their friends steal every weekend." The intensity of stealing lobsters from the lobstermen's traps occurs during the busy summer months when there is a lot of traffic along the coastal waterways. And having lived on or near the ocean for most of my life, I can attest to the fact that poaching does happen and that it is a serious problem for the lobster fishermen.

"Escape Vents"

While great importance has been previously assigned to the purpose of the minimum and maximum gauge laws and other conservation measures, there are other happenings going on in the lobster's life that might well keep it from ever attaining the size of a legal minimum size lobster. We have seen that lobsters that are not of legal size are often kept confined in traps, are hauled to the surface time and time again, and are subject to the handling by the lobster fishermen time and time again. This being not enough, these sub-legals are tossed back into the ocean again and again to make their way to the bottom once, twice, and probably more times until they become *"recruits"* and can be retained by the fishermen.

The rather high rate of mortality and the high incidence of *"cull lobsters"* has moved lobster marine scientists to determine a course of action that would minimize the destruction of lobsters confined to lobster traps <u>*between trap hauls*</u>. They were also interested in reducing injury and possibly death associated with the continuous hauling, handling, and throwing back of sub-legal size lobsters. They, of course, realize that many lobstermen do not haul all of their traps every day, and that many of these traps experience as many as two, three, and even more *"set-over days."*

The resolution of this problem came about through the introduction of a device that would allow most sub-legal size lobsters to escape the trap, while at the same time preventing legal size lobsters from having the same opportunity of doing so. The invention introduced was appropriately called the *"escape vent."* Maine's first escape vent regulation became law in 1979 and has been revised several times over the years, the latest revision occurring in 1999.

The high rate of injuries, deformities, and even mortality in lobsters contained in the traps of lobstermen has been reduced by the incorporation of these escape vents that are located near the bottom section of the parlor compartment. While a good number of sub-legal size lobsters *("shorts")* stand a good chance of leaving a trap before it is hauled to the surface, those sub-legals upon <u>just entering the trap</u> before it is hauled will not be afforded the same opportunity. While the use of escape vents has contributed to the escape and preservation of these smaller lobsters, there are other factors that we have touched upon, and which will be later reviewed, that contribute to injuries, deformities, and mortality of the American lobster.

Catch And Seawater Temperatures

The temperature of the seawater appears to have a great influence upon the growth, mobility, and overall activity of the American lobster. As the seawater temperature lowers, the more the lobster will have the tendency to go off its food, move about less, and seek out a life of seclusion. Conversely, as the seawater temperature rises, the more the lobster will move about, seek out food, molt, mate, and reproduce. This will also be the time when the lobster will be in a much better condition to undergo a molt and the mating process that normally follows when the female of the specie is still in a soft-shell condition. Francis H. Herrick reflected on this in 1909:

> "Whether there is a direct reflex in the lobster to the warming of the shores in the spring or not – it is a fact that it shows a marked tendency, as we have seen, to move shoalward at that time. Further, without any doubt, there is a certain optimum temperature, under the influence of which, when other conditions are favorable, growth is more rapid, and these processes of reproduction and exuviation are most accelerated.
> "The data available, however, do not enable us to determine this point with much accuracy. From these facts, we may infer that the optimum temperature for the lobster lies between 50° and 60° F."

While Herrick's inference that "the optimum temperature for the lobster lies between 50° and 60° F," it is unsure what seawater temperature data was available to him at that time, nor from what locations these measurements were made. Since 1940, however, seawater temperature readings have been routinely taken on a daily basis, with the exception of 1948 and 1949, at the Maine Department of Marine Resources Research Laboratory at West Boothbay Harbor. The annual mean seawater temperatures have ranged from a low of 35.1° F in February to a high of 60.8° F in August (see accompanying table).

A ten-year study of the Maine lobster fishery, undertaken by the author, indicated that from 1972-1981 the first six months of the year were accompanied by a fewer number of pounds of lobsters being landed (approximately 15%) as compared with the number of pounds of lobsters landed during the last six months of the year (approximately 85%). In attempting to link the much lower lobster *"catch"* with seawater temperatures during the first six months of the year, we must be mindful that most of Maine's lobstermen have hauled their boats out of the water for several of these months, and as such, they are not contributing to the landing of lobsters in the State of Maine.

It is under usual circumstances that most lobstermen experience a poor catch when they first start to set and haul their traps early in the spring. They blame it on the cooler than normal temperature of the water at that time of the year. Recently retired lobsterman Arnold Nickerson III of Cape Porpoise, Maine related that "the fishing in the spring of 1994, for example, resulted in a poorer than usual catch because of this cool water temperature – but the lobsters did come in surprising numbers as the year progressed, and especially during the fall months." In trying to explain a declining catch in Massachusetts in 1993, Bill Adler of the Massachusetts Lobstermen's Association wrote, "There's always a number of reasons but one of the biggest is the cold water. It was very cold late in the spring this year and that was true all the way to Canada." Bruce Estrella of the Massachusetts Department of Marine Fisheries stated that "cold water has become a major factor in the depressed lobster catch over the past two years. It's not just local. It goes all the way to Canada. Molting is off as much as 50% when the water remains cold. Lobstermen are heavily dependent on molting locally to increase the number of lobsters of legal size."

These comments and other ramifications of the lobster fishery would seem to suggest that a good season of lobster fishing will probably happen, but the lobstermen have to wait out the water seawater temperatures in order to start gathering in the harvest. This includes waiting out the *"shedders"* of the population that must complete their molts before they start entering traps in search for food.

Using research data published by the Maine Department of Marine Fisheries, it is possible to target the *best and the worse months* for lobster landings in Maine coastal waters and which appears to apply equally as well to the lobster landings in the other major lobster-producing states. The data was formulated by studying the average monthly lobster landings and the average monthly mean surface water temperatures over the ten-year study period of 1972-1981. The following table clearly illustrates that the lobster landings *("catch")* generally increased as the water temperature increased, and that the landings fall off drastically as the water temperature decreased:

LOBSTER CATCH BY MONTH			
(Maine Lobster Fishery – 1972-1981)			
Lobstering Month	Ranking Best Month To Worst Month	Percent Of Total Catch (%)	Mean Surface Temperature (°F)
September	1	19.3	57.6
August	2	18.0	60.8

October	3	17.2	52.1
November	4	12.2	47.6
July	5	11.2	59.7
December	6	6.6	41.8
May	7	5.3	47.8
June	8	4.6	54.3
January	9	2.0	37.3
April	10	2.0	40.6
February	11	0.8	35.1
March	12	0.8	36.4

Catch And Effort

The collection of *"catch"* data (the number of pounds of legal size lobsters landed) and *"effort"* data (the number of traps fished) varies from state to state and is usually compiled from more than one source. The federal government is very much involved in lobster research and the compilation and publication of data relative to the American lobster. The prime agency responsible for discharging these functions is the Atlantic States Marine Fisheries Commission (ASMFC), an important wing of the National Marine Fisheries Service (NMFS), which is a department of the National Oceanographic And Atmospheric Administration (NOAA).

In 1994 there was a rather dramatic increase in American lobster landings in some of the seven major lobster-producing states, especially Maine and

Massachusetts, while four of the seven lobster-producing states experienced a rather significant reduction in lobster landings.

The last year for which complete data is available was in 1998 when approximately 80 million pounds of American lobsters were landed. Maine lobster fishermen accounted for a huge landing of approximately 47 million pounds, and Massachusetts followed with approximately 13 million pounds. The contribution to the total catch by the other five lobster-producing states combined was approximately 20 million pounds.

[Author's Note: The basic raw data used in the following Table (Total U.S. American Lobster Landings (Pounds) – 1998) was published by the Atlantic States Marine Fisheries Commission (ASMFC) in their voluminous report entitled, *"American Lobster Stock Assessment & Peer Review Document - March 2000."* This data was then utilized by the author to formulate certain specifics of importance regarding the lobster fishery, such as: 1) Landings in Pounds and Percentages by State, 2) Landings in Pounds by Month – From the Best Month to the Worse Month, 3) Average Number of Traps Fished Per Boat, and 4) Average Landings Per Trap Fished.

This analysis *excludes* the contribution to the American lobster fishery by the four most southern lobster-producing states of Delaware, Maryland, Virginia, and North Carolina. Data regarding landings and dollars for the American lobster were not always available for these four states, and landings of the American lobster by fishermen working out of these states paled when compared to the other seven lobster-producing states.]

American Lobster Landings and Percentages by State - 1998		
State	**Landings (Pounds)**	**Percent of Total (%)**
Maine	47,036,836	58.7
Massachusetts	13,278,726	16.6
New York	8,525,590	10.7

Rhode Island	5,618,440	7.0
Connecticut	3,715,310	4.6
New Hampshire	1,194,653	1.5
New Jersey	721,811	0.9
7 State Total	80,091,366	100.0

One striking observation that can be made from the data above is that in 1998 Maine and Massachusetts accounted for more than 75% of the total poundage of American lobsters landed in seven of the eleven lobster-producing states. The contribution to the fishery by the four southern-most lobster-producing states, Delaware, Maryland, Virginia, and North Carolina, reflected a meager contribution to the total fishery, and has not been included in this analysis.

While the data above indicate the American lobster landings by state, it may be of interest to know just what part of the vast ocean that these lobsters came from. The Atlantic States Marine Fisheries Commission (ASMFC) has established three specific *"Statistical Stock Areas"* for American lobster habitat. They are 1) Gulf of Maine and South (GOM), 2) Georges Bank and South (GBS), and 3) South of Cape Cod and Long Island Sound (SCCLIS).

The map provided illustrates the magnitude of how far three of the four NMFS statistical stock areas extend out into the Atlantic Ocean - thousands and thousands of square miles. The oft-quoted term, *"bigger is not necessarily better,"* might well apply to these *"Statistical Stock Areas"* when comparing the 1998 American lobster landings from the three basic stock areas. Based on data published by the National

Marine Fisheries Service, the second largest stock area (in size) is the Gulf of Maine and South (GOM), which accounted for an impressive 70.9% of the total landings of American lobster in 1998. The next largest stock area (in size) is the Georges Bank and South (GBS) which contributed 10.6% of the total lobster landings, while the smallest stock area (in size), South of Cape Cod and Long Island Sound (SCCLIS), accounted for 18.5% of the total landings of the American lobster in 1998.

The table, Lobster Catch By Month for the Maine Lobster Fishery – 1972-1981, focused on the largest lobster-producing state of the American lobster - Maine. That table illustrated the *"catch"* (landings) of American lobster by month, the catch from "best month to worse month," the monthly percents of the total landings, and a correlation between catch and mean surface temperature.

The following table, using data published by the Atlantic States Marine Fisheries Commission, has been prepared for the purpose of answering the often asked question: "What are the best months and the worse months for harvesting of the American lobster?" Working with the numbers published by the ASMFC, it is possible to shed some light on this question:

U.S. American Lobster Landings

Best Months To Worse Months

1998

Monthly Ranking By State

Ranking (Best)	Maine	Other Six States*
1st	August	August
2nd	September	September
3rd	October	October
4th	July	July
5th	November	November
6th	December	December
7th	May	June
8th	June	May
9th	April	January
10th	January	April
11th	February	March
12th	March	February

* Delaware, Maryland, Virginia, and North Carolina are excluded.

Up to this point on the subject of *"Catch" Vs. "Effort"* the emphasis has been placed on the *"catch"* in terms of lobster landings. While this is perhaps interesting to be familiar with, probably of more importance to those seeking information about the lobster has to do with the *"effort"* side of the equation. For while it might be interesting to know about annual lobster landings, it is likewise, and perhaps more meaningful, to also have some sort of measure as to how much *"effort"* goes into making these lobster landings?

Fishing effort can be looked upon in many different ways, and by doing so, one can then begin to compare the catch (yield) realized with the effort (boats, traps, fishermen, etc.) that goes into it. For example, during a given year, say 1998, what was the fishing effort expended per boat in producing the landings of American lobster in terms of both poundage and dollar value? What was the effort in terms of boats and the number of traps fished? Because so much emphasis has been placed over the years on the controversial subject of *"overfishing,"* we might further address the question of "what is the average *"yield"* of lobsters per trap fished ? "These matters and questions will be explored and, in the process, other interesting information for the reader may surface as well.

It is at this juncture, however, that one must be again reminded that the data used in making these determinations are estimates at best and that the data were derived from many sources such as lobstermen's log books, landings information, dealer inputs, trawl surveys, and the like. Such determinations are not based on an exact science. Indeed, they are based upon a collection of various data, a "potpourri" of information, that is accessible to the researchers of the lobster, are the best that there is available, and in the overall, should be considered appropriate for use in quantifying catch and effort hypotheses.

The research data that is used in the following analyses is principally that made available by the Atlantic States Marine Fisheries Commission (ASMFC) in its latest Year 2000 assessment of the U.S. American lobster fishery. Much of the focus will be placed on the Maine and Massachusetts lobster fisheries because the data applicable to these two lobster fisheries appear to be more complete, comprehensive, significant, and up-to-date as of the turn into the 21st Century.

So, let us begin with some analyses!

American Lobster Landings			
Seven Major Lobster-Producing States - 1998			
Pounds Landed (lbs)		Dollar Value ($)	
State	(Millions)	(Millions)	Price Per Pound
Maine	47.037	137.189	2.95
Massachusetts	13.279	45.591	3.45
New York	8.826	29.856	3.40
Rhode Island	5.618	20.013	3.60
Connecticut	3.715	12.129	3.30
New Hampshire	1.195	4.702	3.95
New Jersey	.722	2.633	3.65

The Tale of the Lobster

Notes:	• Pounds Landed and Dollar Value Per The Atlantic States Marine Fisheries Commission • Pounds Rounded To The Nearest 1000 Pounds • Dollars Rounded To The Nearest $1,000 • Price Per Pound Rounded To The Nearest 5 Cents

Maine's contribution to the U.S. American lobster fishery is by far large the largest, yet the dollar value and price per pound received by its lobstermen in 1998 were atypical and out of sync with the other six major lobster-producing states. There are some possibilities to consider that might have caused this unbalance. It is within the realm of possibility that the data published by the National Marine Fisheries Service (NMFS) was skewed, or that the Maine lobster fishery was for some unexplainable reason beset with a dramatic drop in price paid to their lobster fishermen. The first possibility can be argued pro or con; and there is another possibility of price failure brought about by the imports of American lobster being trucked in from over the Canadian border. Perhaps also is the possibility that the scores of lobster *"pounds"* held large quantities of lobsters that were carry-overs from the cold winter months. Whatever the reasons, there have been talks with Maine lobster fishermen and others that allude to the fact that many lobstermen were so distressed by the deflated prices received for their lobsters, that they periodically stopped hauling their gear or were on the verge of doing so. As can be seen from the data provided above, average price per pound received by the other six major lobster-producing states was well into the $3.00 to $4.00 price range, and one state, New Hampshire, according to the NMFS, came close to brushing the $4.00 with an average price of $3.95 per pound. And, New Hampshire borders Maine!

Another gauge of fishing effort can be stated in terms of total poundage of lobsters landed as compared with the number of boats used in trapping these lobsters. Based upon information published by the Maine Department of Marine Resources (DMR), an estimated 4,500 lobster boats contributed to the catch of approximately 47 million pounds of lobsters, or approximately 10,500 pounds of legal size lobsters per boat, and with an approximate value of $137 million as reported by the ASMFC. This data applies to the Maine lobster fishery only, but much of the methodology may well apply to the other six lobster-producing states as well. A word of caution, however, is to be interjected at this point: it cannot be assumed that 4,500 boats also represents only 4,500 licensed lobster fishermen, for it is a well-known reality that many lobster boats utilize a "sternman" and that many boats fish so many traps that there are other licensed lobster fishermen involved aboard the boats. There are also many boats that employ licensed *"apprentices"* who are disallowed by law to fish lobster traps in their own behalf; nor does the reported number of boats necessarily reflect the number of boats operated by licensed *"student"* lobster fishermen.

While a breakout of 1998 Maine Lobster Licenses issued is provided at a later point in this chapter, one has to wonder why the State of Maine, or any other state for that matter, cannot provide an accurate account of the number of lobster boats fishing out of its ports. An answer in search of this question appears to be quite simple: assign a person to drive into every port and count the boats tied up at their moorings, one at a time. When would be a good time to count the boats? In Maine waters the appropriate time would be on Sundays during the months of June through August when lobster fishing is prohibited on Sundays. One would, however, have to get a handle on the few boats that are out on the ocean just moving their gear around, which is legal, and the number of boats that are out-of-port and far out into the ocean in the *"offshore"* deep water lobster fishery. The latter situation should not be over-burdensome since many offshore lobster fishermen relocate their gear into the *"inshore"* lobster fishery areas when the seawater temperature warms in late spring and early summer and when even the larger deep water lobsters often have more of a tendency to move a little more shoreward.

One final and interesting inspection of catch and effort is through *"Catch" Vs. "Trap"* analysis. Again it will suffice to focus on the largest lobster-producing state - Maine. According to the data published by the ASMFC, an estimated 2¾+ million traps were involved in Maine lobster landings in 1998. Some 4,500 boats fishing about 625 traps per boat landed approximately 47 million pounds of American lobsters with an approximate value of $137 million. That is a huge number of traps being fished in Maine waters.

So, what might be the average catch of legal size lobsters per trap attributed to the State of Maine during 1998? The answer works out to be <u>approximately 17 pounds per trap per year</u>. And what might be the annual dollar income of one lobster fisherman fishing on one boat in 1998? The answer is approximately $31,000, and is predicated on the following numbers:

Estimated Number of Average Traps Per Boat	625
Pounds Of Legal size Lobsters Landed Per Trap Per Year	17
1998 Published Price Per Pound (Maine Only)	$2.95
Estimated Annual Dollar Value Per Boat Per Fisherman (Rounded To Nearest $1,000)	$31,000

Now, we well know that not every lobster fisherman has an annual income of $31,000. Many probably take home less than this amount, and there is no question that there are many that do, indeed, far exceed this annual income. You don't have an expensive boat, the large fishing-related investments and expenses, an expensive home, two vehicles, all the rest of the costs of living, savings, etc., on a gross income of $31,000 a year! For those interested in some of the philosophical and monetary aspects of lobster fishing as a business, reference to these matters are briefly stated at the conclusion of Chapter 5, *"The Lobster Fisherman."*

Lobster Migration

Periodically during the last one hundred years or so, marine scientists working with lobsters have attempted to determine the extent to which these crustaceans migrate. The pioneer of lobster migration research in the United States was Herman Bumpus who tagged and released about 500 mature female lobsters in the waters near Woods Hole, Massachusetts in 1898. And within four years, Professor A. Mead and L. Williams tagged and released about 400 lobsters in the waters of Narragansett Bay near Wickford, Rhode Island. It was not until the 1950's that lobster migration research was resumed in the United States by lobster marine scientists affiliated with federal, state, and academic research establishments.

Lobster migration research studies undertaken prior to 1965, however, were hampered by a very high *tag loss*, which occurred from the loss of the tagging device during a lobster's molting process. If a tagged lobster molted between the time of release and the time of recapture, the tag would be cast off with the shell of the molted lobster, and the history pertaining to that particular lobster would be lost as well. However, a breakthrough in lobster tagging technique was put into place in 1965 with the introduction of the *"sphyrion tag,"* sometimes referred to as the *"spaghetti tag."* The sphyrion tag is a three-element device consisting of a stainless steel anchoring mechanism, a connecting thread, and the tag itself. The anchoring device is driven into the muscle of the lobster at the point of juncture of the shell of the carapace and the shell of the tail section (abdomen). Connected to the anchoring device is a thin polyethylene thread that is attached to the other end and to a supple PVC (polyvinyl chloride) tube. The tube, measuring a mere 7mm in diameter and 55mm in length, identifies and reveals specific information about each tagged lobster. If properly implanted, the tag should remain in the muscle of the lobster, even if the lobster should happen to undergo a molt.

Research studies in more recent years have also incorporated the use of the *"rock lobster tag,"* which is similarly anchored into the muscle between the shell of the carapace and the shell of the tail section. The rock lobster tag differed slightly with the sphyrion tag and consisted of an oblong-shaped piece of plastic that measured approximately 3mm in diameter and 10mm in length.

Another tagging device has often been used as a *"back-up tag,"* referred to as a *"cinch-up tag,"* and is used by researchers who are interested in evaluating specific happenings that <u>could</u> take place with a tagged lobster. They first want to know the rate of success, or failure, of either the sphyrion tag or the rock lobster tag; they also want to know if a tagged lobster has molted since the time of its last release. The cinch-up tag is normally attached very snugly around either the ripper claw or the carpus appendage of the ripper claw. If a lobster pulled to the surface in a trap has a cinch-up tag, but does not have a sphyrion or rock lobster tag, that tells the researchers that something went wrong with their method of anchoring the tag device into the muscle of the lobster. Also, with a cinch-up tag being still in place, it would tell the researchers that the lobster has not molted since its last release. However, if the cinch-up tag is missing but either the sphyrion tag or the rock lobster tag is still in place, then that would tell them that the lobster had most likely molted since its last release.

The information contained on the tagging device enables the lobster marine scientist to track each released and recaptured lobster. Where did it go? How far did it travel and in how much time? In what direction did it travel? Has the lobster molted since its last release? These are the main questions that the marine scientist is looking for answers to when undertaking lobster migration research.

Unfortunately, these marine scientists have never gotten a total return of information from the lobsters that they have tagged and released into the ocean. One of the major reasons for this is that there is always going to be a certain number of lobstermen who elect not to participate in any study, and regardless of who or what parties are carrying out the study. Rather than provide essential information on a *"Migration Study Report Card,"* they simply retain any legal size lobster as part of their catch and do not return the requested information. Then there are the dragger fishermen who drag their heavy nets along the bottom in their pursuit of groundfish. Any tagged lobsters caught up in their nets would most likely be retained as part of their "catch" and not reported on as such.. Then there are the tagged lobsters that end up in *"ghost traps"* or are set upon and injured or mortally wounded by other marine species occupying the same ocean habitat. All of these possible occurrences equate to the fact that these tagged lobsters will never be reported on and the information that might have been revealed will be lost to the marine scientists doing lobster migration research studies.

From the migration research conducted over the years, it is a fact that the overwhelming majority of the inshore lobster population *do not* migrate significant distances, and that these lobsters are recaptured within very close proximity to the points from where they were last released.

One such study was conducted by Jay S. Krouse and others of the Maine Department of Marine Resources and supports the contention that there *is not* a mass migration of Maine's inshore lobster population. The findings of the study (1975-1977) do, however, indicate that a very insignificant number of tagged lobsters did stray considerable distances from their release points and that a very few of them did indeed *"walk"* incredible distances. In 1975, some 2,882 legal size lobsters were released from three locations off the Maine coast: Kennebunkport, Boothbay Harbor, and Jonesport. Four months after their release, 65% had been recaptured, and by the time the study was concluded 2½ years later, about 76% of the tagged lobsters had been recaptured. Of these recaptures, 88% were removed from traps that were within five miles of their release sites. There were only eight lobsters that moved more than fifteen miles from their release sites, and two of these eight lobsters traveled remarkable distances. One was a male with a carapace length of 3½ inches that migrated a distance of 63 miles in 1 year and 4 days before being hauled up in a lobster trap off Boston, Massachusetts. The other was a female lobster with a carapace of 3½ inches that managed to migrate 185 nautical miles in about six months before being hauled up in a trap off Tiverton, Rhode Island. That lobster most likely took a *"short cut"* through the Cape Cod Canal!

While there appears to be little or no large trend in the outward migration of Maine's inshore lobster population, there is some movement of the lobsters within the confines of most inshore fishing areas. Adult lobsters have the general tendency to move somewhat *"seaward"* in the late fall and *"shoalward"* in the late spring. The availability of food and other environmental conditions have a great deal to do with whether a lobster will move – or whether a lobster will stay put. Well documented criteria such as the need to molt, the laying or hatching of the female's eggs, and seawater temperature are all part of the biological and environmental conditions that will urge the lobster to move on or remain relatively in the same location. Lobster marine scientists generally believe that when an adequate supply of food is available, most of the total inshore lobster population will remain pretty much where they are even though the seawater temperature continues to fall lower and lower during the cold winter months. In addition to the conditions mentioned above, the lobster will most likely remain where it is if there is a generally rocky bottom, rock crevices, mud, or any form of substrate that it can use to hibernate and conceal itself from its natural enemies.

Lobster migration studies in Maine have shown that the inshore lobsters that do move tend to do so in a shoalward direction or along the coast in a west to southwesterly direction from where these lobsters were tagged and released. The movement of inshore lobsters in other directions has been found to be minimal.

For the large size *"offshore"* lobster fishery, it's another story. These deep-water lobsters exhibit a much greater need or desire to migrate and travel longer distances than do their *"inshore"* counterparts. Dr. Richard A. Cooper of the National Oceanographic and Atmospheric Administration at Woods Hole, Massachusetts suggests that every year in the spring and early summer, about 30-35% of the deep-water lobsters undertake various stages of migration from off Cape Cod, the Georges Bank, the Gulf of Maine, and other offshore lobster habitats. Cooper explains this migration as follows:

> "They undertake various stages of migration… to get into shallow warm water… or a lot of them come into Cape Cod or Rhode Island Sound. And they are doing it obviously to get into warm water for the females to shed and for the males to mate with the females. These lobsters… start coming around the tip of Cape

> Cod in waters 30 to 100 feet deep along a very narrow slope that occurs only a quarter of a mile off the tip of Cape Cod.
>
> "They're walking on the tips of their walking legs and their fantail sections open up, and they move right along at a pretty good clip. They will frequently stick their large claws into the sand along a slope if the current is in the wrong direction, and they'll spend five or six hours there if the tide is running pretty good and until the tide changes again."

During an informal conversation with Dr. Cooper and a few Maine lobstermen at an Annual Maine Fishermen' Forum at Rockland, Maine, he further described the migration of these offshore lobsters:

> "When the offshore migration starts, it starts at the southern end of the range, and that almost the entire lobster fishery along Southern Long Island Sound is big lobsters in deep water. That migration starts in April or May. And when you get up off Rhode Island, the population of _each canyon_ starts its migration in June. This is followed by the lobster population of Georges Bank, which starts its movement in July or August. It is estimated that about thirty to fifty percent move."

Dr. Cooper and other marine scientists are able to make these observations about lobster behavior in the wild by *"trawl surveys,"* SCUBA diving or by going to the bottom of the ocean in research *"submersibles."* They are of the opinion that some of the offshore lobsters undertake round-trip migrations of 300-400 miles, and they have observed that the outer parts of their slender walking legs are badly eroded from walking these large distances on the ocean floor. Lobsters, of course, do little *"swimming"* in the true sense of the word; rather they make the best use of their four pairs of walking legs along with occasional help from their tailfans and large crusher and ripper claws.

Another aspect to consider is just how far a tagged lobster might travel, if it travels at all, before getting hauled up in a lobsterman's trap. This, of course, depends upon the fishing *"effort"* that is going on in close proximity to where a lobster is released. The greater the number of traps being fished, the greater is the possibility that the lobster will be trapped almost immediately. Lobster marine scientists suggest that the movement of the larger size lobsters is considerably more than the smaller of the specie. In Thomas's words, "what I'm suggesting... is that movement is dependent on size."

Another important lobster migration study undertaken by the Maine Department of Marine Resources commenced in the spring of 1977 and involved only *"juvenile"* sub-legal size lobsters. The Maine DMR was interested in the movement of the smaller lobsters, their growth, their molting frequency, mortality, and the like. During the reporting period of 1979 through September of 1982, only 831 of the 4,740 tagged lobsters had been recaptured in traps or by SCUBA diving. This small recapture rate (17½%) appears to support the opinion of Dr. Cooper and others that these very small lobsters are inclined to hide and seek out a life of almost total seclusion in rock crevices, mud burrows, or any artifact that exists on the ocean floor.

The low recapture rate might also be due to the possibility that some of these lobsters had grown following a molt, which they certainly do, and had advanced from being sub-legal size lobsters upon their last release to being lobsters that now just barely meet the minimum carapace length, often referred to as *"recruits,"* that would allow them to be classified as legal size lobsters. If a lobster fisherman decided to keep such a lobster and not report on it, then the information pertaining to that lobster would never be retrieved for use by the researchers studying lobster migration. However, of the 247 legal size lobsters that were returned to the ocean and reported on by cooperating lobstermen during the 2½ year study period, 142 (57%) of them had moved less than one mile from the release site in the Sheepscot River, 93 of them (38%) had traveled a distance of one to four miles, and the remaining 12 (5%) traveled a distance of five to eight miles from their release points. From this research study data, Jay S. Krouse suggested that:

> "Lobsters found up river at the release point were usually caught during the summer and early fall when catches are generally highest in shoaler water of the river, whereas lobsters that moved down river in a southerly direction to relatively deep waters were caught throughout the season: particularly in the spring-early summer and late fall-winter when fishermen concentrate their effort in greater water depths where lobster catchability is higher due to warmer water temperatures."

Jay S. Krouse also had a very important point to make with regard to these tagged lobsters: "Contrary to what one might expect, the elapsed time between release and recapture did not affect how far the animals moved. Lobsters at large for 2-3 years moved no further than those free a year or less."

One of the most ambitious lobster migration studies, and probably the most controversial, ever conducted in Maine waters commenced in 1983, continued throughout 1984, and progressed into 1985. This particular study was originally requested by the Maine Lobstermen's Association and stemmed from their concern about the movement of large size "*female lobsters*" in Maine waters. Were these lobsters migrating and contributing to the offshore fishery? Were these lobsters great travelers and were they legging it northward toward Nova Scotia, southwesterly toward New Hampshire, or even more southerly in the direction of Massachusetts, the Cape Cod Canal or around the tip of Cape Cod and into Long Island Sound and points south? Or were these larger size lobsters staying relatively close inshore like the smaller size lobsters that had been tagged and researched in past lobster migration studies?

Acting upon request of the MLA, a joint undertaking was entered into by the MLA, the Maine Department of Marine Resources (DMR), and the University of Maine at Orono (UMO). They struck an agreement whereby the MLA would concentrate their effort on soliciting the support and cooperation of the Maine lobstermen, the DMR would procure, tag, and arrange for the release of the tagged lobsters, and the UMO would build a computer system to track and report on sundry bits of information pertaining to the tagged lobsters.

In October of 1983, 1,145 healthy and vigorous large size female lobsters with a carapace length of 3½ inches or more were rounded up from lobster dealers, lobster pounds, and lobstermen in the Stonington area. They were tagged, V-Notched, and released in deep water depths ranging from 190 to 250 feet. The Maine DMR and UMO college assistants then headed south to Boothbay Harbor where 1,230 more of the larger size female lobsters were procured, tagged, V-Notched, and released in 150 to 240 feet of water just south of Damariscove Island. The main tagging device was either a sphyrion or rock lobster tag, and as a back-up measure, a cinch-up tag was snugly tightened around the carpus appendage of one of the large claws of each lobster.

The lobster migration study was continued in earnest in 1984, but was accompanied by a few changes in methodology. A major problem was encountered when there were not enough of the large size female lobsters to be found readily available in the Stonington and Boothbay Harbor areas. The Maine DMR, therefore, made a decision to use a mix of both male and female lobsters for its 1984 tagging operation. It was at this point that the Maine DMR appeared to have made an error in judgment: *they decided to V-Notch the male lobsters as well as the female lobsters,* a practice that was unheard of in the Maine lobster fishery. The DMR and assistants from the UMO used approximately 4,000 tags to tag nearly 2,000 lobsters. But during the process of the tagging, they appeared to have made another error in judgment and apparently threw caution to the wind: *they decided to remove the first pair of swimmerets (the "gonopods," or "stylets") from each of the 890 male lobsters!* But this unusual procedure did not get by the Maine lobstermen who carefully inspect each and every lobster they remove from their traps. They must know, as we have seen, if a lobster is an egg-bearing female, a V-Notched female, or a lobster of either sex that might meet the maximum legal size requirement. And the Maine lobstermen were not fooled! In fact, they were very upset about the V-Notching of the male lobsters, and they were even more upset about the removal of the gonopods from these male lobsters. Edward Blackmore, then president of the MLA apparently tried to smooth over the problem and maintain peace by rationalizing that the *"alteration"* was done on lobsters that would have been purchased from lobstermen, dealers, and pound owners and, as such, would have been headed for the dinner table anyway, had they not been purchased for the tagging research project. However, because of the V-Notching and gonopod alteration that had been applied to the male lobsters, there were many lobstermen who went on record as saying that every one of those male lobsters caught in their traps would go into their catch - rather than returning them to the ocean and reporting on them.

While the success of this particular lobster migration study was somewhat hampered by the lack of cooperation and participation on the part of some lobstermen, enough data was derived to form some trends regarding the movement of lobsters. The results from the lobsters obtained from the Boothbay Harbor area indicated that the majority of recaptures (94%) were trapped within 20 miles (an average of 3.4 miles) from their release sites, and only 6% of the recaptures traveled greater distances, an average of 60 miles. A somewhat similar migratory dispersal occurred with the lobsters from the Stonington area where approximately 85% of the recaptures were trapped within 20 miles (an average of 7.1 miles) from their release points, and approximately 15% of the recaptures traveled greater distances, an average of 64.4 miles.

The loss of some of Maine's lobster population to other states, such as New Hampshire and Massachusetts appears to happen, but not to a significant degree. The lobster, however, knows no boundaries, and the V-Notch cut into the tailfan of sexually mature female lobsters *"gives them away"* as being a part of the Maine lobster fishery. Lobster marine scientist Bruce T. Estrella of the Massachusetts

Department of Marine Fisheries acknowledges the Maine to Massachusetts lobster migration:

> "There is additional evidence of infiltration of migrating lobsters from outside Massachusetts territorial waters. Numerous observations of lobsters marked with V-shaped notches… have been made during the study period off Cape Ann and outer Cape Cod. These lobsters are presumed to have migrated from Maine waters where a V-Notched female protection program is enforced. This is an industry-sponsored program intended to protect a segment of the breed stock off the coast of Maine… V-Notched females were generally large, averaging greater than 100mm CL.
>
> "Large migrant lobster (> 100mm CL, many of which were oviferous females) from either offshore or Canadian and Maine waters were more numerous off Cape Ann and outer Cape Cod than in other Massachusetts coastal regions sampled. Geography may be a factor in concentrating these lobsters, since these two regions are adjacent to steeply gradients which lead to a much greater depth range than is available to other coastal regions sampled."
>
> [Author's Note: CL = Carapace Length; 100mm = 3.94 inches; Oviferous = Carrying ova (eggs)]

The results of these lobster migration studies can be summarized as follows:

- Certain lobsters amongst both the inshore and offshore populations do migrate.

- Lobsters that do migrate from Maine's inshore fishery have a tendency to travel along the shore and in a southwesterly direction.

- A few of Maine's lobsters might travel great distances, but tagging studies reveal that the vast majority of them do not move about to any great extent and are recaptured within close proximity from where they were released.

- Probably the greatest migration takes place in the offshore fishery where larger size lobsters abound in greater numbers. These lobsters live in a deep-water habitat such as in and around the numerous *"canyons"* making up the North Atlantic Continental Shelf of the United States.

- There is a tendency for some lobsters to move toward shallow waters when the coastal waters warm and toward deeper water when the coastal waters cool.

- New Hampshire and Massachusetts appear to be the main beneficiaries of those lobsters that do migrate from Maine and Canadian waters.

These assessments of American lobster migration appear to be in agreement with the findings and conclusions of the Atlantic States Marine Fisheries Commission in its March 2000 publication of the *"American Lobster Stock Assessment & Peer Review Document:"*

> *"There are varying hypotheses as to whether larger undergo directed migration, temperature-mediated movement, or no migration at all Haakonsen and Anoruo 1994). Seasonal inshore-offshore movements of adult lobsters have been demonstrated by tagging studies conducted in the Gulf of Maine (Cooper, et. al., 1975), east of Cape Cod (Estrella and Morrissey 1977), and in southern New England (Saila and Flowers 1968, Cooper and Uzmann 1971, 1977, Briggs 1985, Cobb et al. 1989).*
>
> *"In general, migrating lobsters move offshore in the fall and winter and inshore in the spring and summer. For ovigerous females, such behavior exposes eggs to warmer temperatures, thus enhancing egg development."*

For many years the lobstermen and the commercial *"draggermen"* have been at odds with one another regarding the right to co-exist and fish for lobsters in the deep water off the New England coastline. The issue has gained more attention in recent years when the lobstermen and their associations have banded together to put up a concerted effort to stop the draggermen from taking lobsters from the offshore fishing grounds.

The offshore lobstermen point their fingers at the draggermen for damaging their gear and for the loss of untold numbers of traps because of their dragging of heavy nets along the bottom of the ocean floor. The deep-water lobstermen also charge that these dragger fishermen are netting huge quantities of the larger size lobsters, while the dragger fishermen consider them to be a legitimate bi-product of their normal catch of groundfish such as cod, haddock, flounder, sole, and redfish *("ocean perch")*. On a long list of complaints, the lobstermen further contend that the trawls and rakes that are used to sweep along the ocean bottom are causing injury and sometimes death to countless numbers of lobsters, and that such injuries contribute to the high incidence of *"cull lobsters"* (a lobster with either one or both of its large claws missing). They also complain that shell damage is making some lobsters prone to shell disease, such as the *"red tail" disease ("gaffkemia"),* and the like. And if all of this isn't bad enough, damaged and un-buoyed lobster traps contribute to the mortality of these lobsters entombed in *"ghost traps"*

The dragger fishermen, however, seem to hold an entirely different viewpoint concerning this gear conflict. From their perspective they claim that the damage to the lobstermen's gear is probably minimal and even at that, they are free to catch lobsters in their nets or by any other means at their disposal. They take the position that they operate far out in deep water and beyond the three-mile limit that extends into federally controlled waters. This, in effect, disallows any state agency from being involved in the offshore lobster fishery.

An article concerning this subject reported on some of the dialogue that took place during a meeting of the England Fishery Management Council in March of 1995:

> "...the Council also voted to issue permits that will allow lobstermen to catch the crustaceans between three and 200 miles off the coast of New England... Lobstermen generally are licensed by their home states.
> "Lobstermen are appealing for some restrictions on draggers who they say are stripping the ocean floor of their catch. The lobstermen say boats displaced from the Georges Bank by the groundfishing ban there are taking huge lobster catches.
> "Lobstermen using traditional traps also say egg-bearing females are not being returned to the sea as required by law.
> "Lobstermen who attended the meeting, which attracted an overflow crowd of more than 100, expressed fears about the future of their livelihood. 'But I'm doing it because I've got a wife and two kids,' said David Marciano, 29, of Beverly.
> "Maine representatives have argued without success for years for a total ban on dragging for lobsters in all New England waters. David Cousens, president of the Maine Lobstermen's Association, said two Gloucester draggers landed lobster catches of 9,000 pounds and 8,500 pounds last week."

Another issue of complaint by the offshore lobstermen is the flouting of the law that prohibits the *"scrubbing"* of eggs off any female lobster that clearly shows the presence of an egg mass (ova), and that this is being practiced by offshore commercial fishermen in order to bring these lobsters ashore and sell them as marketable lobsters. Virtually all lobster fishermen have held the long-standing belief that the large male and female lobsters must be protected, and they are convinced that the large female lobsters represent the present and future brood stock of the fishery. And, as the law dictates, at least in Maine, all *"berried"* female lobsters must be returned to the ocean.

One glaring report of *"scrubbing"* occurred in the spring of 1996 when a 72-foot dragger returned to Gloucester, Massachusetts with 6,400 pounds of lobsters. The dragger was approaching Gloucester Harbor when it was apprehended and boarded by personnel of the Coast Guard and Massachusetts Environmental Police. Among the catch were 110 female lobsters that showed signs that eggs had been removed from the underside of their abdominal tail sections. There was further evidence of lobster eggs on the dragger's deck. In an article appearing in the Boston Herald on June 10, 1996, Jules Crittenden reported on the sentiments of one individual, Howard Nickerson of the Offshore Mariners Association, who said, "I don't like the idea of scrubbing lobster and I don't know of any fisherman who does... it's a black mark for everybody all around... it's poor judgment on the part of the fisherman and the federal government..."

While the owner of a commercial fishing vessel stands to lose his license, face thousands of dollars in fines, and could be jailed for scrubbing offenses, the real hurt will be felt by the lobster fishermen in the months and years to follow. We can make a supposition to prove this point: The eggs from 110 female lobsters will never be given the chance to be hatched out on the bottom of the ocean. However, if they were able to do so, and if approximately 0.1% of these larval lobsters managed to survive the perils on or near the surface of the ocean, then approximately 5,500 of them would return to the bottom of the ocean again to continue their lives as tiny *"lobsterlings."*

Assuming: 110 *"berried"* lobsters

Assuming: 50,000 eggs per lobster

= 5,500,000 eggs hatched

X 0.1% survival rate

= 5,500 surviving lobsters (out of the potential of 5,500,000!)

It is perhaps no wonder that lobstermen sometimes look upon the draggermen with anger and resentment. They claim that the draggermen have damaged their trap gear, have exploited the lobster resource by dragging the bottom and retaining lobsters as a by-catch, and are sometimes caught in the act of scrubbing the eggs from berried female lobsters!

The plight of the dragger fishermen, however, has no doubt been brought about by their declining catches of groundfish. Then to make matters even worse, the federal government has placed severe restrictions, in the form of quotas, on <u>how many</u> pounds of certain species of groundfish that the draggermen may catch and bring into port. The federal government has also placed restraints as to <u>where</u> the draggermen can fish. Hundreds of square miles of ocean that were traditionally exploited for the groundfish there are now closed off to the commercial dragger fishermen in order to allow the necessary time for the groundfish population to become replenished.

Until such a period of time is realized, the rift between the offshore lobstermen and the offshore draggermen will most likely continue in earnest. Perhaps the gear conflict issue could become a non-issue, or an issue of little importance, if the two parties could keep their distance. This would be a situation whereby the commercial dragger fishermen would stay clear of those waters that are clearly marked by the presence of lobster buoys, and by the lobstermen making an attempt to set their strings of traps in locations where the draggermen do not drag the bottom for groundfish.

A Maine law does, however, place restrictions on the trawling, seining, and netting of lobsters. According to the *"Compilation of Lobster Laws and Related Regulations"* published by the Maine Department of Marine Resources, the following restrictions apply:

> "It shall be unlawful to fish for or to take lobsters by use of an otter or beam trawl, a scallop drag or trawl, seine or net or to have in possession any lobsters, regardless of their source, on board any boat rigged for otter or beam trawling, scallop dragging or trawling, seining or netting.
>
> **Exceptions:**
> 1. No violation of this section shall occur if the lobster is immediately liberated alive in the coastal waters.
> 2. This section shall not apply to any boat rigged for otter or beam trawling, scallop dragging or trawling, or seining if all nets and scallop drags are removed from the boat.
> 3. This section shall not apply to any boat rigged for netting if there are no finfish taken by gill net aboard."

In reviewing the several hundred pages of text in the March 2000 *"American Lobster Stock Assessment & Peer Review Document,"* prepared by *the Atlantic States Marine Fisheries Commission,* nary a word was mentioned about the landings of the American lobster – or the possible impact on the fishery – by commercial dragger fishermen. In querying the absence of this item of consideration, it apparently was considered as not being of pertinence to the assessment and not a matter to be taken up in discussions and review. Perhaps it was looked upon as being more of a state lobster fishery issue or an issue of little consequence to the overall. Or perhaps the non-inclusion was of the mindset: *"let's not meddle with it."*

Trap Tags

The State of Maine has a *"Lobster Tag System"* in place that requires the incorporation of a trap tag for each trap being fished in Maine waters. The tags, which must be purchased (at about 20 cents each), usually on an annual basis, from the Commissioner of the Maine Department of Marine Resources, identify ownership. The tags, in addition to ownership identification, are used to keep track of the number of traps being fished by Maine lobster fishermen, and this, of course, is for the purpose of keeping tabs on the trap limit restrictions that prevail in the Maine lobster fishery. As is also the case for Massachusetts, there appears to be no one requirement regarding the information printed on each trap tag,, nor is there any consistent usage of tag color. While each trap tag must bear the license number of the license holder, some tags also include the imprinting of the license holder's last name, and the tag color can been observed as being black, orange, blue, red, and the like. The trap tag in common usage is of hard polyvinyl construction, and is usually looped around one or more of the cross sections located in the *"parlor"* compartment of a wood or wire lobster trap.

Maine and Massachusetts are very stringent regarding the concept of ownership, and further require that buoys, crates, and *"lobster cars"* be identified in a similar manner. This is no doubt of necessity to the functions of marine law enforcement personnel who must have and seek proof of gear ownership and to readily identify situations where there is a reason to suspect that the *"poaching'* of lobsters or lobster gear is happening along the coastal waterways.

So, in essence, trap tags serve to prove ownership on the one hand, while on the other they provide a means of attempting to make the lobstermen abide by the trap limits as stipulated by state law. This latter goal is just another in a series of steps in the attempt to control the intensity of the lobster fishing that is occurring in coastal waters.

Trap And Trawl Limits

Certain restrictions in lobster trap usage by each lobster fisherman are set by law. Maine's inshore lobster fishery, as we have seen, is divided into seven *"Zones,"* six of the seven zones of which has a trap limit of 800 traps per licensed fisherman; the seventh zone, *Zone E,* is located in and around the Boothbay Harbor area where lobster fishing is very intensive. The trap limit for this area is 600 traps.

In addition to the trap limits allowed for commercial lobster fishermen, Maine also has a non-commercial license available with a limit of five traps, and for the younger set who are full-time students and under the age of 23, there are licenses available that allow for the fishing of 150 traps. In Maine, as in Massachusetts, there is a non-commercial license available that carries with it a limit of only five traps that may be fished.

Maine has a few areas where the number of traps fished on trawls is restricted. Notable among these are 1) the Casco Bay Area where only 12 traps are allowed per trawl, 2) the Sheepscot Bay and Sequin Island Area where only 3 traps per trawl are permitted, 3) the Linekin Bay Area where only 2 traps are allowed per trawl, and 4) the Kittery Area where only 10 traps per trawl are allowed.

The trap limits and trawl limits as set forth by Maine law are so established for obvious reasons: intensive fishing for lobsters by so many lobstermen operating out of so many boats is oftentimes being done in rather tight geographical ocean areas. This can be envisioned by anyone looking out onto the ocean on any fine morning during the late spring, summer, and fall months - there is a lot of lobster fishing going on out there - and oftentimes scores of lobster boats can be seen from the shoreline as their skippers move from trap to trap in search of the lobster.

It might also be mentioned that Massachusetts has its own trap limit restrictions. The present law allows each commercial lobster vessel to fish a maximum number of 800 traps. There is also a seasonal commercial lobster permit issued to qualified full-time students that allows each licensee to take and sell lobsters from June 15th to September 15th. Such a permit restricts the licensee to the fishing of only twenty-five traps. In Massachusetts there is also a non-commercial permit available for the taking of lobsters and crabs for personal use. Personal use in this case applies to the permit holder and members of the holder's immediate family that reside in the same residence. The trap limit for such license holders is set at ten traps only.

Trap Construction

It may be presumed that the reason for stipulating certain specifications for lobster traps is that without set specifications there would be lobster traps of all shapes and sizes sitting on the bottom of the ocean.

Lobster traps in use today are commonly of the vinyl-coated wire type, but there are still many of the older wood traps still being fished. Regardless of the type of trap material used, both Maine and Massachusetts laws dictate that the volume of a lobster trap be a maximum of 22,950 cubic inches. The largest of the traps used by the inshore lobstermen are the three-compartment traps consisting of a *"kitchen"* compartment and two *"parlor"* compartments; the smaller two-compartment traps are constructed with one *"kitchen"* compartment and one *"parlor"* compartment.

All traps, as discussed previously, feature one, but usually two *"escape vents"* that are located near the bottom of the parlor compartment, and each trap must have a *"biodegradable panel"* that breaks down and self-destructs over time in order to release lobsters stranded in un-buoyed *"ghost traps."* The biodegradable panel is located at the top of the trap's parlor compartment. Finally, the incorporation of a *"trap tag"* is secured in place on the top portion of the trap's parlor compartment and is utilized for the purpose of tracking lobster traps being fished, trap limit contingencies, and to provide proof of ownership.

Trap Molestation And Removal

Marine laws in effect in Maine and Massachusetts, at the very least, address two areas of concern that affect both the lobster fishermen and the possible *"unaccountability"* of lobster gear. In Maine, the law is very explicit and may be considered as being rather harsh, but as is often the case, the end is justified by the means. For the Maine lobster fishery, the Maine Department of Marine Resources in its publication, *"Compilation of Lobster Laws and Related Regulations,"* sets forth the following legal restrictions and ramifications for possible offenses:

Molesting Lobster Gear

"No person may raise, lift, transfer, possess or in any manner molest any lobster trap, warp, buoy or car except as provided in this section.

1. **Permitted activities.** Lobster traps, warps, buoys, and cars may be raised, lifted, transferred, possessed or otherwise molested by the following:

    a. A marine patrol officer;

    b. The licensed owner;

    c. Any person having written permission from the licensed owner; and

    d. Any person authorized by rule pursuant to subsection 2.

2. **Promulgation of rules required.** The commissioner shall promulgate rules ... authorizing the removal of traps, warps, buoys or cars that are

washed up above the mean low tide mark or are otherwise abandoned or lost.

3. **Prohibition.** Traps, warps, buoys or cars may not be used for fishing by any person other than the licensed owner unless with written permission from the licensed owner.

4. **Additional penalty.** If the holder of a lobster and crab fishing license violates this section by cutting a lobster trap line, the court shall:

    a. Order that person to pay to the owner of the trap line that was cut an amount equal to twice the replacement value of all traps lost as a result of that cutting…"

Lobster Trap Removal

"…Any person who possesses traps, warps, buoys or cars and is not a Marine Patrol Officer, the licensed owner or someone with written permission from the licensed owner or a Marine Patrol Officer, shall be in violation of 12 M.R.S.A. S6434."

The Massachusetts law relating to lobster gear molestation or removal is not as specific or detailed as compared with Maine laws. In the *"Massachusetts Division of Marine Fisheries, Abstract 2000,* the following wording is used:

> "It is illegal for anyone, except the owner, to handle, destroy or molest any lobster fishing gear, including any gear swept under the shore, beaches or flats whether public or private…"

It appears that the message to be imparted from the above-stated laws is two-fold:

1. All boaters, lobstermen and pleasure-crafters alike, should never ever cut or molest the lobster gear of others, and,

2. Lobster gear (traps, warp, buoys, etc.), regardless of their condition, should not be removed by the "passer-by," but only by the owner or with permission from the owner or other authority. This applies especially to wood and wire traps that have been cast up on rocks and beaches along the coastline. This occurrence is common and is usually part of the aftermath resulting from violent storms such as hurricanes and "northeasters."

Mortality And Disease

Whether intentional or not, there are certain practices and conditions present that weigh into the picture of mortality of the American lobster. Some are induced by the acts of man, some are biological in nature, and some are strictly environmental. Without question, man is the main cause of death of the lobster, and this is often referred to as *"fishing mortality."* Lobster fishermen, after all, are the main predators of lobsters, as can be evidenced by statistical data regarding the landings of lobsters in 1998. This is basically brought about through harvesting of lobsters by thousands of fishermen in the United States and the Canadian Maritime Provinces in the inshore and offshore lobster habitats. But a distinction should be made regarding the harvesting of lobsters as an act of "normal occurrence" during the conduct of the business, and the *"premature"* injury and possible death brought about by other practices and conditions as has been referenced previously in this chapter.

There are several dominant causes that can be linked to the *"natural mortality"* of lobsters:

1. Predation
2. Disease
3. Environmental Factors
    - Violent storms
    - Toxic metals
    - Chemical spills.

Lobsters are often subject to attack by their natural predators before, during, and after the molting process, but the greatest risk of being attacked occurs when the lobster is in a soft-shelled condition following a molt. Underwater research has also revealed that *"lobsterlings"* (very small lobsters) probably suffer the highest loss by bottom-dwelling marine life that share the same habitat. Predation decreases as the lobster grows larger and when there is a favorable availability of shelter (under rocks, in rock crevices, in mud burrows, and in artifacts resting on the ocean floor.)

Lobster marine scientists hold the opinion that disease in the lobster is more of a contributory factor in death and injury than is the case for predation and that mortality as a result of disease can impact lobsters of any size. Parasitic organisms are a chief cause of mortality and certain of these organisms, especially a type of marine worm, can lead to the mortality of ova in *"berried"* female lobsters.

Lobsters with shell damage or in a weakened state fall common prey to these marine organism (parasites, etc.) The rate of shell infection is also higher with ovigerous female lobsters because they molt less frequently and therefore, over time, accumulate an abundance of organisms that become fastened to their thick and hard shell structure. This is especially so for the *"carapace"* and the *"tail section"* of the female lobster.

Shell disease can also occur outside of the lobster's natural habitat such as in *"lobster pounds,"* lobster *"pools,"* and even in the lobstermen's *"lobster cars."* A high incidence of shell disease is common among lobsters being held in confinement in lobster pounds while awaiting shipment throughout the region, the country, and the world. Lobsters being held in this manner and in a restricted space over a long period of time, and with a poor exchange of seawater, are very prone to acquiring shell disease and the bacterial infection that usually follows. It was reported in an edition of the *"Journal of Shellfish Research"* that:

> "Periodic outbreaks of shell disease in Nova Scotia and Maine coastal impoundments have historically affected lobsters imported into Massachusetts and other states.
> "Complaints have been voiced by local lobster dealers of unaesthetic appearance, weakness, and enhanced mortality among these imports. Heavy organic loading and poor water quality in impoundments, which allow bacteria to flourish, appear to be responsible."

One common type of shell disease is known as the *"red tail disease,"* generally referred to by lobster marine scientists as *"gaffkemia,"* and which can inflict widespread mortality to lobsters being held in captivity.

Another aspect of disease and resulting mortality in the American lobster is called the *"gas disease."* The presence of air leaks in the piping or tubing of a lobster holding tank over an extended period of time can bring about the mortality of lobsters therein confined. The leaking of air within the system will, in time, produce an unsatisfactory level of oxygen; there will result a super-saturation of the water, all of which is toxic and oftentimes fatal to lobsters.

A third type of disease and resulting death in the American lobster is becoming better known and more frequent in the last of the 20th Century and the beginning of the 21st Century. This negative effect on the lobster habitat is brought about chiefly as a result of the actions of man and sometimes occurring as *"accidents"* in which *"chemical spills"* are spewed into the ocean as pollutants and causing either death or a disdainful appearance in the American lobster. Oil spills involving oil tankers, tugs, and barges have become suspect as the cause of death in the American lobster, and the discharge of chemically-treated sewage water is the suspected culprit that has caused blanched-out lobsters to be pulled up in traps and have caused lobsters to suddenly disappear from the scene in some of the most productive lobster habitats. A very significant case in point has recently occurred in Long Island Sound where a large number of lobsters were consistently pulled up in traps and were pronounced dead-on-arrival when being hauled aboard in what has been termed as *"the lobster kill-off"* in Long Island Sound. And occurring at about the same time was another incidence in the mortality of lobsters as lobstermen fishing their traditional waters in Salem Sound were pulling up lobsters that were bleached-out and unattractive for human consumption. In the first instance, the suspected culprit was oil spills into the Sound, while in the latter situation the finger is being pointed at the huge sewage treatment plant at Salem Willows, Massachusetts as being the possible contaminant through the release of chlorine and other chemicals from the outflow pipe that extends quite some distance out into the ocean from the sewage treatment facility. In both types of situations, the lobster marine scientists and state lawmakers are still in the process of defining the problems and with no short-term resolution in sight.

One final threat to the American lobster is the possibility of exposure to toxic metals while being held in confinement. Certain common metals such as copper and brass are extremely toxic to lobsters. Technological research and innovations have brought about the practice of storing lobsters in tanks that employ the use of vinyl tubing and piping through which seawater is constantly re-circulated. By taking these precautions against toxic metals, the mortality rate of lobsters being held in captivity in lobster pools and restaurants has been found to be very minimal.

Summary Of The Major Causes Of Mortality In Lobsters

Death Induced Or Caused By The Actions Of Man

- Capture of Legal size Lobsters by Lobster Fishermen
- Capture of Lobsters of All Sizes by Draggermen
- Capture of Lobsters of All Sizes by SCUBA Divers

- Capture of Lobsters of All Sizes by Pleasure Boaters *("Poaching")*
- Careless Handling of Sub-Legal Size Lobsters by Lobster Fishermen and Their Helpers
- Careless Handling of Legal size Lobsters by Others Along the Entire Distribution Chain
- Indiscriminate Dragging of the Ocean Bottom by Draggermen
- Indiscriminate Release of Cut-Off Trap Warp by Pleasure Boaters
- Accidentally-Induced Pollution (Oil Spills/Sewage Treatment Discharge).

Death Induced Or Caused By The Environment

- Severe Coastal Storms
- Ghost Traps
- Cannibalism
- Disease
- Pollutants
- Inadequate Salinity Levels.

Death Induced Or Caused By Natural Causes

- Predation by Other Forms of Marine Life
- Molting Difficulties
- Old Age.

Licensing

It should come as no surprise that the licensing of those people who fish for the American lobster is fairly rigid, and is extremely so in the State of Maine. There is a myriad of licenses available for those that qualify, much of which has been discussed previously in this chapter. Maine has a complex variety of licenses consisting of ten major categories that apply to a universe of people, including people in their younger years, folks in their later years, students,, and those that are involved in an "apprentice program." And there are licenses available for a few people who want to set out a few traps for non-commercial fishing for the American lobster.

For those interested in the licensing of the lobster fishery in the State of Maine, the following table sets forth the parameters and annual fees set forth by the Maine Department of Marine Resources in its 1998 publication, *"Compilation of Lobster Laws and Related Regulations:"*

TYPE LICENSE		ANNUAL FEE ($)
LOBSTER AND CRAB HARVESTING		
Class 1	Commercial	93.00
	Promotion Surcharge (a)	25.00
Class 1	Commercial, Under Age 18	46.00
Class 1	Commercial, Over Age 70	46.00
Class 2	Commercial	186.00

		Promotion Surcharge (a)	50.00
Class 3		Commercial	279.00
		Promotion Surcharge (a)	75.00
Student		Full-Time, Under Age 23 (b)	46.00
Non-Commercial (c)			46.00
APPRENTICE LOBSTER AND CRAB HARVESTING			
Apprentice		Under Age 19 (d)	46.00
Apprentice		Age 18 And Above (d)	93.00
Apprentice		Age 70 And Above (d)	46.00
a. A portion of the Promotion Fee goes toward the operation of the "Lobster Fund."			
b. Allows only a 150 trap limit			
c. Allows only a 5 trap limit			
d. Disallows fishing lobster traps in his or her own behalf. Apprentices must have a sponsor.			

In addition, there are higher annual fees applicable to businesses that qualify for wholesale and retail dealer licenses, for the transport of Maine lobsters, and those businesses that deal in lobster meat.

The "Lobster Fund" previously alluded to "is used for the purpose of lobster biology research, of propagation of lobsters by liberating seed lobsters and female lobsters in Maine coastal waters and of establishing and supporting lobster hatcheries. The commissioner shall give priority to purchasing seed lobsters and may provide purchased seed lobsters to lobster hatcheries. The remaining seed and all female lobsters shall be liberated in the coastal waters after V-Notching them in the right flipper. The commissioner may authorize the expenditure of money in the Lobster Fund for research and development programs which address the restoration, development or consideration of lobster resources. The commissioner may authorize the expenditure of money in the Lobster Fund to cover the initial costs of developing and delivering the educational component of the apprentice program..." (Per the Maine DMR 1998 publication, *"Compilation of Lobster Laws and Related Regulations".*)

A significant portion of the annual Lobster Fund surcharges goes towards the augmentation of license processing activities and the operation of the Maine Lobster Promotion Council, the goals and purposes of which have been discussed earlier in this chapter.

The Commonwealth of Massachusetts, the second largest lobster-producing state, also has a great variety of lobster fishing licenses, some of which are summarized as follows:

TYPE LICENSE	ANNUAL FEE ($)	
	RESIDENT	**NON-RESIDENT**
Coastal Lobster Permit	260.00	N/A
Offshore Lobster Permit/+99 (a)	260.00	520.00

Seasonal Lobster Permit (b)	65.00	130.00
Non-Commercial Lobster Permit (c)	40.00	60.00
a. A Federal Lobster Permit is required		
b. Allows the taking and selling of lobsters to a licensed dealer from June 15th to September 15th. 25-trap limit. Issued to full-time students only		
c. For the taking of lobsters for personal use only. 10-trap limit.		

Massachusetts also levies annual fees for a variety of other type of licenses, including annual fees for boats, wholesale and retail dealers, trucks, bait dealers, and so forth. A complete listing of annual license fees, restrictions, and the like, is available from the Massachusetts Division of Marine Fisheries. The remaining lobster-producing states also have definitive licensing requirements and restrictions, but the examination of the Maine and Massachusetts data are considered sufficient to prove the point that the lobster fishery is well regulated and its laws are very comprehensive.

Law Enforcement

The Maine Bureau of Marine Patrol is the oldest law enforcement agency in the state and is, without a doubt, more extensive and far-reaching than is the case for any other lobster-producing state. This is so because of the sheer number of personnel and resources needed to carry out the functions of the Bureau.

The bureau is headed up by a Chief, a Deputy Chief, two Division Managers, and a Superior Officer for each of the three Districts. The bureau is further supplemented by a Special Services Section that is commanded by a Superior Officer, has two mechanics, and is equipped with a seaplane.

Split into two Divisions and three Sections, there are 34 Marine Patrol Officers who are assigned to strategic locations along Maine's seemingly endless miles of coastline. Each Marine Patrol Officer is assigned to cover a certain area of the ocean that extends from Kittery in the south near the New Hampshire border to Lubec near the Canadian border. They discharge their duties in rather small 21' Boston Whalers and much larger boats that are up to 44' in length, some of which are specially equipped and capable of operating at very fast speeds. The larger boats are used primarily for carrying out the policing of law enforcement laws and regulations pertaining to the far offshore commercial bottom-fishing boats. All boats, the seaplane, and boat resources are owned and maintained by the Maine Bureau of Marine Patrol.

The time expended and the people resources available are divided among the lobster fishery, the commercial deep-water dragger and trawler fishery, and the bureau's boating safety training programs. The estimated allocation of the Bureau's available resources and responsibilities are set forth in the following table:

Enforcement of Lobster Rules and Regulations	30%
Enforcement of Groundfish Rules and Regulations	55%
Carrying Out of the Bureau's Boating Safety Programs and Bureau Training Activities	15%

The Marine Patrol Officers have sweeping powers in the discharge of their responsibilities. A few of the excerpts taken from the 1998 publication of the Maine Department of Marine Resources "Compilation of Lobster Laws and Related Regulations" pamphlet is indicative of just how far these sweeping powers are as they apply to each of the 34 Marine Patrol Officers that operate within the areas of Maine's lobster fishery:

- Applicants for the position of marine patrol officer who qualify under the officers' code and pass the examination… may be appointed by the commissioner to hold office…
- Officers shall enforce all marine resources' laws and may arrest and prosecute all violators.
- They shall have jurisdiction and authority in all areas where the laws for which they have responsibility apply… in addition to their specific powers and duties, marine patrol officers are vested with the authority to enforce all laws of the State and may arrest for violations of any criminal laws.
- Any marine patrol officer, in uniform, may search without a warrant and examine any watercraft, aircraft, conveyance, vehicle, box, bag, locker, trap, crate or other receptacle or container for any marine organism when he has probable cause to believe that any marine organism taken, possessed, or transported contrary to law is concealed thereon or therein.

What are some of the infractions of the marine laws that the Marine Patrol Officers are on the lookout for in Maine coastal waters? The infractions, virtually all of which have been addressed earlier in this book, will be presented briefly for the purpose of understanding of the major violations: The summary that follows applies to the Maine lobster fishery only:

The Marine Patrol Officers have the authority to investigate:

- The proper licensing of all people in the lobster fishery
- The proper identification of the numbering of lobster boats, lobster gear, vehicles, and other aspects of the fishery
- The possibility of taking lobsters of illegal size, such as the minimum and maximum *"carapace length" sizes* allowed by law
- The possibility of taking female lobsters showing eggs under the back shell of the tail section
- The illegal scraping or removable of the eggs showing on a female lobster
- The possibility of taking female lobsters with a *"V-Notch"* cut into one of the appendages of the tailfan
- The possibility of the practice of *"poaching"* (stealing) of lobsters or trap gear from a lobsterman
- Etc.

The one major complaint, it seems, among some lobster fishermen is that they rarely see a Marine Patrol Officer. The major reason for this appears to be the case of, "Do you want to keep a closer eye on the 'good guys' or the 'bad guys'?" The Maine Bureau of Marine Patrol allegedly takes the position that its primary concerns are keeping abreast of the activities of the "bad guys" who violate Maine's marine laws pertaining to the marine fisheries. And that would appear to be both logical and justifiable if the lobster fishery is to be protected and preserved.

Chapter 7. "The Feast" – Buying, Cooking, And Eating The Lobster

During one of many visits to the Maine Department of Marine Resources Laboratory at West Boothbay Harbor, a scribbled-out verse was noticed hanging from the wall in front of the desk of marine scientist Glenn Nutting. The verse, written by his little daughter, Beth, read as follows:

"Lobsters are

fun to eat,

I like to pick

them up.

Wons I put

a lobster on the

floor and my dog grold

at it."

LOBSTERMEN'S "EAT-IN-THE-ROUGH" COOPERATIVE, BOOTHBAY HARBOR, MAINE

CAPE PORPOISE LOBSTER CO. INC., CAPE PORPOISE, MAINE

This chapter will focus on the buying of lobsters, preparing them for the table, and the enjoyment of the "feast" that is sure to follow. The menus and the lobster preparation instructions provided at the conclusion of the chapter will hopefully be of invaluable assistance to the consumer who wishes to enjoy dining on the lobster at home rather than at a restaurant or other seafood and lobster establishments.

The lobster fisherman is keenly aware at all times that trends in lobster prices usually follow a fairly predictable annual pattern. Lobsters caught during and even after the *"shedding"* (molting) period will have shells that feel somewhat rubbery and softer to the touch than is the case for hard-shelled lobsters. In addition, the soft-shelled lobsters will contain more body fluids and the meat will be less firm when compared to the hard-shelled lobsters. The lobsterman also realizes that the price he receives for his lobsters will be dependent upon the supply and demand of the marketplace. Oftentimes the lobsterman will hold off and store his catch until such time that the price per pound is more favorable for him. While the impounding of lobsters in this manner seems to make sense, many lobstermen find it impractical to store their lobsters for a long period of time. A lobsterman might have a cash-flow problem and needs to turn his lobsters over on pretty much a daily basis. He might also not have a *"live car"* available for storing his lobsters, and he knows that lobsters held in storage for a long period of time are often subject to a certain amount of disease and mortality. For these and other reasons, many lobstermen will sell their lobsters on a weekly or daily basis.

Robert Delano Martin

The Buying Of Lobsters

Royal (Roy) Bellveau, a long-time employee of the James Hook Lobster Company in Boston, has been tinkering around with lobsters for practically his entire lifetime, and he has many amusing stories to tell about his experiences over the years. Most of his stories relate to the way retail customers perceive and look upon lobsters and how they react to them. He lamented, "One of the biggest fallacies going is that people think that when lobsters are out of the water, they're going to smother. You close the box up, and for some of these people, the damndest look comes over their face – like you're closing the lid on the coffin!" He goes on to say that many a retail customer will punch holes in the box before they put their lobsters in the refrigerator. Claiming that this is unnecessary, Bellveau adds, "Lobsters have gone from here to London or someplace without any holes in the box. But you can't convince people of that. Lobsters are good for 24 hours after they die, so don't throw them away." He tells the story about a telephone call he received from a customer in California who called to inform him that two of the twelve lobsters arrived dead on arrival, so she threw all twelve lobsters away. She thought they were poisonous or something and she wanted her money back!

Sometimes people complain about an occasional lobster that has *"droopy claws."* In most places these lobsters are referred to as *"sleepers,"* and many people will not eat those lobsters. The truth of the matter is that there is nothing to fear in eating a lobster with droopy claws. Regarding the taste of a droopy-clawed lobster, Bellveau insists that "after picking the meat out, I defy anyone to tell the difference."

Concerning the physical handling of a lobster that is about to be dropped into the pot, Bellveau had this to say:

> "People are so afraid of the lobster. Put them in a bag, go home, put in the water, dump them in... then they say they cover them up so they won't have to listen to them. It's not true. I've boiled and seen more lobsters than people have seen in a lifetime."

What this expert lobster handler is saying is that lobsters don't cry, sing, gasp, or emit all sorts of sounds when they are placed in boiling water. One might simply dismiss such a notion as "an old wives tale."

The physical handling of the lobster is both easy and safe. Most perfect lobsters are purchased with both large claws closed shut with either a rubber band or a peg, and there is no chance of injury when the animal is handled by the *"carapace"* (the head and body section). The lobster's crusher and ripper claws are able to come back just so far and are incapable of extending their reach far back over the carapace when held by the hand. For the lobster to extend its claws back any further could result in that lobster *"throwing"* or *"shooting"* its claws, which is very unlikely. So if the lobsterman, the wholesaler, or the retailer has banded or pegged the claws shut, the consumer should have no fear at handling such a lobster. For those who still possess a fear of the animal, fear not and simply grasp the lobster by the carapace and everything will be all right.

Greater affluence of the fishermen and improvements in the techniques for storing and transporting lobsters have placed the animal at our very doorsteps. And whether they are *"summer lobsters"* or *"winter lobsters,"* the prices paid for them over the years have gradually increased in response to the many factors that influence supply and demand.

When purchasing lobsters directly from lobstermen, the dealers have usually sort them out by size, weight, and condition. The sizing and grading is not always uniform and might very well vary from state to state and from one dealer to another. However, the grading and classification of lobsters is fairly well established, as set forth in a table published by the Maine Lobster Promotion Council:

Classification of Lobsters	
1 to 1-1/8 lb	Chickens
1¼ lb	Quarters
1½ to 1¾ lb	Halves
1¾ to 2½ lb	Selects
2½ to 3½ lb	Large Selects
Over 3½ lb	Jumbos

While there is a great demand for the lobster during the annual tourist season, much of Maine's lobsters are sold and shipped to out of state dealers and to consumers who purchase them on an order placed over the telephone or an order placed over the "internet." Many purveyors of the lobster will sell their lobsters by far fewer categories than is illustrated in the above table. Many outlets will grade their lobsters as *"chickens," "mediums," "selects," jumbos," and "culls,"* while at a large supermarket, lobsters might be sold only as *"chickens"* or *"selects."* A

The Tale of the Lobster

general rule of thumb is that the price that one pays for a live lobster will be in direct proportion to the size of that lobster - the larger and heavier the lobster, the higher will be the price. The exception to this probability would be a lower price for a *"cull"* lobster (one claw missing) or a *"jumbo size"* lobster that is very heavy and very huge in size.

The dramatic increase in live lobster prices over the years can best be illustrated by looking at prevailing prices at specific points in time. In 1903 live lobsters sold for only 25 cents a pound at a fish market in Tilton, New Hampshire. In the 1940's live chicken lobsters could be purchased at the Beverly (Massachusetts) Fish and Lobster Company for 69 cents a pound. Moving ahead into the 1970's and beyond, a dramatic increase in lobster prices was experienced. In the 1970's, lobsters were selling for a little over $3.00 per pound; in the 1980's, they were selling at about $4.00 to $5.00 per pound, depending on their availability. Jumping ahead to January of 1995, live lobsters were being sold for about $5.50 to $7.00 per pound at Rowand Fisheries in Beverly, Massachusetts:

Grade	Price Per Pound
Chickens	$5.49
Mediums	$6.49
Selects	$6.99

Lobsters purchased at supermarkets will generally cost the consumer more money than when purchased from a local lobster pool or fish market. An example of this was the price of lobsters charged by Shaw's Supermarket in Beverly, Massachusetts in January of 1995: $5.00 per pound for chicken lobsters and $7.49 per pound for select lobsters. Not too far distant, the Atlantic Lobster Company in Saugus, Massachusetts was advertising their jumbo size lobsters (5-8 pounds) for only $5.95 per pound.

It should be pointed out, however, that the higher price of live lobsters purchased during the late fall to late spring months occurs when the supply is depressed, as compared with live lobsters purchased when the supply is greater. This is attributable, as we have seen, to those circumstances whereby the ocean water has warmed, the lobsters are moving about on the ocean floor, are in search of food, and end up more often in the lobsterman's traps.

On the basis of this supply and demand relationship, prevailing spot prices during the first week of April, 2000 for chicken size lobsters and lobster meat at typical lobster pools or fish markets in five principal cities of the six major lobster-producing states are provided in the following table:

City	Price Per Pound ($)	
	"Chickens"	Lobster Meat
Portland, ME	7.99	32.95
Portsmouth, NH	6.99	29.00
Boston, MA	6.99	35.00
Narragansett, RI	6.50	n/a

Stamford, CT	6.99	24.00
New York, NY	6.85	n/a

An analysis of the lobster price data presented above indicates that the price per pound of "chicken size" lobsters ranged between $ 6.50 and $ 6.99, with the exception of Portland, Maine where the price was $ 7.99 per pound. A dealer in Stamford, Connecticut sold lobster meat (claws and tails) at the lowest price of $ 24.00 per pound.

As compared with hard-shelled lobsters, the soft-shelled lobsters (*"shedders"*) will contain more water, more body fluids, and the meat will be less firm. A little advice from those that know about such matters would suggest that the purchaser of live lobsters be selective at the time of their purchase by requesting only lobsters with a hard shell. Should the purchaser be uncertain as to whether every lobster going into the bag or box is a hard-shelled lobster, it is not unreasonable to ask the counter-person for permission to examine each of the lobsters. Perhaps a little repetition at this point is in order: grasp the lobster by the shell of the carapace and press into the sides of the carapace with the fingers. If the shell of the carapace is soft to the touch when squeezed, and is easily depressed by the fingertips, then that lobster is most likely <u>not</u> a hard-shelled lobster; rather, it is a lobster that has not had sufficient time to harden its new shell structure following the molting of its old shell.

There are those people who do not like to handle a live lobster, let along boiling it, steaming it, removing it from the pot, and then go through the somewhat tedious process of prying the meat out of the shell. Nor do they relish the *"mess,"* the odor and the chore of having to dispose of the remains after the *"feast"* is over. For them, there are many purveyors of lobsters that sell cooked lobster meat by body parts, most usually the meat from the claws and the body sections. The question arises: how many lobsters does it take to produce one pound of clear lobster meat? The answer to this question would depend upon the condition of the lobsters and how much the lobsters weigh; but as a general rule, 4-5 hard-shelled 1 to 1-1/8 pound two-claw *"chicken"* lobsters will yield about one pound of clear and solid lobster meat. If the lobsters happen to be one-claw *"cull"* lobsters, perhaps 5-6 lobsters will be required to yield the same amount of lobster meat.

One might hear the claim that there is no finer-tasting lobster than those caught in Canadian waters, and there are still others who make the same claim about the so-called *"Maine Lobster."* Probably there is no one that can really support nor prove either of these contentions, because the taste of the lobster is excellent across its entire range, from Labrador to North Carolina. Then there are those who wouldn't think about buying a *"super jumbo size"* lobster because they are operating under the assumption that the meat is tough, stringy, and less tasty than is the case for a smaller size lobster. Some of those in the business say this isn't so, and that the taste of the meat of the larger size lobster is every bit as good as the meat from a smaller size lobster, and the meat is not tough and stringy – one as compared with the other.

Tomalley, the liver of the lobster, is considered a delicacy by some and can be found in the body section of the lobster. Also to be found there are tiny pockets of sweet and tender meat. The body sections of lobsters can be purchased occasionally and are usually sold for a very minimal price. For those who prefer not to deal with live lobsters at home, most fish markets and lobster pools will cheerfully boil and sell the lobsters at the same price as being charged for live lobsters.

Lobsters that are graded as *"selects"* (roughly 1½ pounders or a little more) ordinarily demand the highest price of all lobsters sold over the counter. *"Culls"* (one-claw lobsters) always carry the lowest price tag when sold to the public. While a cull lobster can contain only one claw, or no claws at all, the *"no-claw"* will most likely <u>never</u> be seen by the consumer; and a lobster in this condition is usually boiled by the dealer and its contents sold as lobster meat. Such a lobster has either *"thrown"* or *"shot"* both of its claws, has lost them in a battle in the wild, or has been mishandled by those in the trade (or a combination of the above), all of which makes the animal very unappealing. Without either of the large claws, and often the long antennae and some legs, such a lobster is a repulsive and ugly-looking animal!

Those consumers who have access to the lobster, and can afford the price to go along with it, will probably have two or three *"feeds of lobsters"* during the summer or early fall months. The general tendency is that most people who do so will eat two, maybe three lobsters without any difficulty. Many a summer tourist enjoying a vacation get-away along the New England coastline will often manage to fit a lobster *"eat-in-the-rough"* experience into the itinerary. Along the coastline there are seafood restaurants and Lobstermen's Cooperatives here that cater to the summer tourist and offer hot boiled lobsters on site. They provide all of the essentials: the lobsters, butter, tables, benches, bibs, picks, shell-crackers, and even a towelette for cleaning the hands and fingers. All one has to do is select the lobsters, pay the piper, and sit down and enjoy the feast. Two well-known caterers to the eat-in-the-rough diner are the Lobstermen's Cooperative at West Boothbay Harbor, Maine and Woodman's on the causeway in Essex, Massachusetts. This same type of eat-in-the-rough experience is also featured as an important part of the *"Homecoming Week"* festivities that are held during the summer in many cities and towns in New England and elsewhere. One day during the week is called *"Lobster Festival Day"* and features boiled lobsters, steamed clams *("steamers")*, and corn-on-the-cob. Twin lobster specials have become increasingly popular in seafood restaurants. Prices range between $13 and $16 for a complete meal that includes rolls, butter, and a choice of vegetable. Prices charged for lobsters at the more up-scale restaurants will be, as might be expected, considerably higher.

Lobsters purchased for cooking at home should be *"live and kicking"* as they are being removed from the seawater in a holding tank. Oftentimes a lobster will flip its tail section back and forth several times while being handled, all of which is an assurance that the lobster is a very healthy one. While there sometimes can be a long lag time between the time a lobster is caught and the time it is sold, there is always the possibility that the consumer might experience a *"sleepy lobster."* A lobster in this condition is characterized by droopy claws and without a great amount of life to it, but experts on the matter attest to the fact that there is simply nothing to be feared, and that such a lobster is good for human consumption for as long as twenty-four hours after its natural mortality. So just because the claws might be drooping and the lobster appears to be sluggish, there should be no reason for not cooking and eating that lobster. Should there be any doubt as to whether or not a listless lobster is *"dead,"* examine the many appendages that make up the mouthparts to determine movement. A live lobster that is *"alive"* will keep its mouthparts in constant motion, especially so when handled or breathed upon. A *"dead"* lobster would not show any indication of movement of these mouthparts.

Another measure of freshness of the lobster is a "self discovery" that is referred to as *"the tail return test."* Whether in its natural habitat on the ocean floor or while in captivity, the lobster generally carries its tail section in a curled-up position. Furthermore, when a live lobster is dropped into a pot of hot boiling water, it will eventually curl its tail under. And this should be the position of the tail section <u>*even after it has been boiled or steamed.*</u> If the tail section returns to its natural curled-under position, there is no problem; if it does not, there is probably something the matter with that lobster.

Cooking The Lobster

There are those among us that evoke very strong emotions when it comes to dropping a live lobster into a pot of water boiling away at 212° F. And shoving a live lobster into the oven brings about the same level of response. Indeed, there are those who cannot bring themselves to do such a thing, and perhaps the chore should be given to a person who has little or no such misgivings, and is willing to do so.

The vast majority of people who partake of lobsters prefer to have the lobsters boiled or steamed. There are fewer in number who will prefer their lobsters baked, and still a fewer number will favor a broiled or fried lobster. With regard to the boiling of lobsters, a precautionary note should be made as to the manner in which a live lobster is dropped into a pot of hot boiling water: a lobster dropped into the pot *tail first* not only prolongs the death of the lobster, but subjects one to being splashed by the boiling water. This would come about as a result of the lobster flexing its strong and muscular tail section back and forth in direct response to the heat. Therefore, as a safeguard against being scalded, a lobster should always be dropped into the pot **head first**. When preparing a lobster for baking or broiling, the lobster should be flipped over onto its backside and cut from head to tail with a sharp knife.

Many people are extremely bothered by the movement of several of the lobster's appendages after the lobster has been subjected to boiling water or slit down the underside with a knife. Movement of the walking legs, antennae, the many mouthparts and other appendages is quite noticeable. However, this movement is attributed to the chemical endings of the nerves that stimulate the muscle tissues and which, in turn, bring about uncontrolled reflexes. During a visit to the Maine Department of Research Laboratory at West Boothbay Harbor, lobster marine scientist Jay S. Krouse dissected a 1¼ pound lobster. He did this part by part and organ by organ, even removing the heart, the brain, the legs, and the antennae, and the various mouthparts still continued to twitch. To put this in a matter of a few words, it's nothing more than reflexes. There might also be an occasion when a blackish substance is found in the body section of the lobster. According to Krouse, this blackish material is brought about by the absorption of the ovaries in the female lobster. This is usually caused by an unsatisfactory environmental or biological circumstance just before the lobster was about to extrude her eggs. In some quarters they are referred to as *"black lobsters,"* but there is certainly no danger to humans in the eating of such a lobster. Simply scrape the blackish material to one side and enjoy the feast!

While there are some publications available that address the preparation and eating of the American lobster, there is really no purpose to go beyond the basic instructions provided by the Maine Department of Marine Resources in its publication, *Recipes From Maine – LOBSTERS."* The author, however, has taken literary freedom by reformatting these instructions in a step-by-step sequence and has modified some of the instructions in order to simplify and hasten understanding. The author has also taken the liberty of adding a few comments here and there, but only when considered necessary and appropriate.

[Author's Note: All of the American lobster is edible, with the exception of the following parts:
- The shell parts and antennae (of course)

- The small *"crop"* or *"craw"* located in the body section

- The dark intestinal vein that runs through the abdomen (tail section).

RECIPE # 1: MAINE BOILED LOBSTER

- ➢ Add about 2-3 inches of water to a pot deep and wide enough to hold the lobsters

- ➢ Add a little salt to the water

- ➢ Bring the water to a boil

- ➢ Add the lobsters

- ➢ Cover the pot immediately

- ➢ Wait for the water to come to a **second boil**
- ➢ Set the timer for 15 minutes (see Author's Note)
- ➢ Remove the lobsters after 15 minutes
- ➢ Drain the water and body fluids by placing the lobsters "face down" in the kitchen sink
- ➢ Serve the lobsters either hot or cool with a cup of hot melted butter or margarine.

[Author's Note: The Maine Department of Marine Resources recommends that lobsters be boiled for 18-20 minutes after the water has come to a second boil. However, after interpreting the boiling instructions of two major dealers who boil lobsters on a daily basis, it appears that this boiling time may be too long for smaller "chicken size" lobsters that weigh about 1 to 1-1/8 pounds and which are the predominant size lobsters purchased by the retail customer.

The following table, therefore, provides boiling for lobsters that weigh from 1 to 2 pounds, in increments, as recommended by a major lobster retailer in Boston. For a 1 to 1-1/8 pound lobster, boil for 15 minutes (after the water has come to a second boil). For lobsters heavier than 1-1/8 pounds each, add 1-½ minutes of boiling time, as follows:

Weight (Pounds)	Boiling Time (Minutes)**
1 to 1-1/8	15
1 ¼	16 ½
1 ½	18
1 ¾	19 ½
2	21
** After the second boil].	

RECIPE #2: MAINE BAKED LOBSTER

- Lay the lobster on its backside
- Cross the large claws
- Hold the claws in place with one hand
- Cut a deep incision with a sharp knife down through the entire length of the lobster
- Remove the intestinal vein and craw
- Stuff the lobster as full as possible with lobster stuffing. (Refer to the recipe below)
- Pour hot melted butter or margarine over the stuffing and sprinkle generously with Parmesan cheese and paprika
- Place large lettuce leaves over the stuffing to preserve moisture while baking
- Bake for 30 to 40 minutes at 350 to 400 degrees F
- Remove from the oven and remove the lettuce leaves
- Serve with a side cup of melted butter or margarine.

Baked Lobster Stuffing (For 4 Lobsters)

- Roll or grind 16 old-fashioned "Common Crackers" (the unsalted variety) into a fine crumb mixture
- Season with salt and pepper
- Mix the crumbs with ¼ cup of hot melted butter or margarine and ¼ cup of whole milk

- ➢ Mix with sherry wine to the proper consistency.

RECIPE # 3 MAINE BROILED LOBSTER

- ➢ Lay the lobster on its backside
- ➢ Cross the large claws
- ➢ Hold the claws in place with one hand
- ➢ Cut a deep incision with a sharp knife through the entire length of the lobster
- ➢ Remove the intestinal vein and craw
- ➢ Spread the dressing generously in the cavity of the lobster (Refer to the recipe below)
- ➢ Place large lettuce leaves over the dressing to preserve moisture while broiling
- ➢ Place the lobster on a broiler rack (About 8-10 inches from the broiler heating element)
- ➢ Broil for 25 to 30 minutes
- ➢ Remove the lobster from the oven and remove the lettuce leaves
- ➢ Serve the lobster with a side cup of hot melted butter or margarine.

Broiled Lobster Dressing (For 4 Lobsters)

- ➢ Mix together 1½ cups of cracker crumbs or cracker meal and ½ teaspoon of salt

- Moisten with 2 tablespoons of Worcestershire Sauce and 1 cup of hot melted butter or margarine.

RECIPE # 4: MAINE LOBSTER SALAD

- Use boiled lobster meat that has been chilled
- Cut the chilled lobster meat into small pieces
- Add 1 teaspoon of mayonnaise or salad dressing for each cup of lobster meat
- Mix thoroughly
- Chill once again
- Place on crisp lettuce leaves
- Sprinkle with paprika
- Serve on a salad plate
- Provide a side dish of mayonnaise or salad dressing.

RECIPE # 5: MAINE'S OWN LOBSTER SALAD ROLL (FOR 4 LOBSTER ROLLS)

- Use lobster meat that has been chilled
- Cut 2 cups of lobster meat into small pieces (see note below)
- Blend the lobster meat with 2 tablespoons of butter or margarine
- Add ¼ cup of fine diced celery, if desired
- Mix well and let stand in the refrigerator until ready to be served
- Split and toast 2 "hamburger rolls" or "hot dog" rolls

- Spread the rolls with butter or margarine, if desired
- Fill the rolls with the lobster mix and serve on a salad plate.

Note: The Maine Department of Marine Resources recipe suggests ½ cup of lobster meat per person.

RECIPE # 6: MAINE LOBSTER STEW

- Boil 4 chicken size lobsters (1 to 1-1/8 pounders) Refer to Recipe # 1 for boiling instructions
- Remove the lobster meat immediately after boiling
- Save the tomalley and coral (if desired)
- Cut the lobster meat into fairly large pieces
- Mix the lobster meat with the tomalley and coral (if desired)
- Cook over a low heat for 10 minutes
- **Add slowly** 1 quart of rich hot milk and **stir constantly**
- Season with salt and pepper
- Allow to stand for **5 to 6 hours** before reheating and serving.

Note The Maine Department of Marine Resources offers the following suggestions for the preparation of Maine Lobster Stew:

"The preparation of the perfect Maine Lobster Stew is simple, but it does not allows for short cuts. Stirring is most important in this recipe, otherwise it will curdle. According to the experts of fine Maine cookery, the important steps to success in creating the perfect Lobster Stew are:
 1. The partial cooking before gently adding the hot milk, a mere trickle at a time

2. The constant stirring until the stew blossoms a rich salmon under your spoon
3. The "aging," since every hour that passes improves the flavor.

"Two days are set by the masters of the Lobster stew for "aging," with a minimum period of 5 or 6 hours, and afterwards the stew should be kept under refrigeration until it is reheated and served."

RECIPE # 7: MAINE TANGY LOBSTER (FOR 4 PEOPLE)

- Place all ingredients (refer to ingredients list below) in the top part of a double boiler
- Melt ingredients over hot water
- Stir and blend well
- Cut the lobster meat into fine pieces
- Add the lobster meat to the blended mixture
- Cook for 6 minutes
- Serve on toast, with lemon wedge, if desired.

Ingredients

- 2 cups of boiled lobster meat (refer to Recipe # 1 for boiling instructions)
- 1/3 cup of butter or margarine
- 1 teaspoon of Worcestershire Sauce
- 1 tablespoon of lemon juice
- 1 teaspoon of dry mustard
- ½ teaspoon of salt
- pepper, to taste.

RECIPE # 8: LOBSTER CASSEROLE (FOR 4 PEOPLE)

Ingredients

- 3 boiled chicken size lobsters (1 to 1-1/8 pounders - Refer to Recipe # 1 for boiling instructions.)
- 2 tablespoons of butter or margarine
- 2 tablespoons of all-purpose flour
- 1 cup of top milk or thin cream
- ½ cup of whole milk
- ½ teaspoon of dry mustard
- 5 slices of bread
- bread crumbs or cereal flakes
- salt and paprika.

Directions

- Melt the butter or margarine in the top part of a double boiler
- Add the flour, mustard and salt
- Make a paste of the mixture
- **Add slowly** the thin cream (or top milk) and the whole milk
- Stir constantly to make a smooth mixture
- Stir and cook until the mixture thickens
- Break the bread crumbs into small pieces (but not too small)
- Add the bread pieces (remove the crust first)

The Tale of the Lobster

- Mix well and pour into a casserole dish
- Top off the mixture with buttered coarse crumbs or cereal flakes
- Bake in the oven at 350 degrees F for a period long enough to reheat the mixture and to lightly toast the topping
- Serve immediately.

RECIPE # 9: MAINE LOBSTER CROQUETTES (FOR 12 CROQUETTES)

- <u>Ingredients</u>
- 3 boiled chicken size lobsters (1 to 1-1/8 pounders) Refer to Recipe # 1 for boiling instructions
- ¼ pound of butter or margarine
- ½ cup of finely chopped onions
- ¼ cup of finely chopped celery
- 1 cup of heavy cream sauce
- ¼ cup of bread crumbs
- 3 eggs
- 1 ounce of sherry wine.

<u>Directions</u>

- Melt the butter or margarine
- Add the chopped onions and chopped celery
- Add the finely chopped lobster meat
- Allow the mixture to simmer for 3 to 4 minutes

- Add the cream, the bread crumbs and the sherry wine
- Chill the mixture until it becomes firm
- Shape the firmed-up mixture into croquettes
- Fry in deep fat at 175 degrees F for 3 to 4 minutes
- Serve immediately.

RECIPE # 10: MAINE LOBSTER COCKTAIL (FOR 1 COCKTAIL)

- Grind ¼ cup of boiled lobster meat (Refer to Recipe # 1 for instructions)
- Blend 2 tablespoons of tomato catsup, 2 tablespoons of sherry wine, 1 teaspoon of lemon juice and 6 drops of Tabasco sauce
- Chill thoroughly
- Serve in a cocktail glass.

Eating The Lobster

With most of the preparation effort accomplished, it's now time to sit down around the dinner table to get at the delicious lobster meat that lies hidden beneath the hard shell of the lobster. The mining-out of the lobster meat might involve a little work, but with the proper working tools the task will become much less of a chore.

First of all, it is strongly recommended that hot lobsters coming out of the pot be placed in the kitchen sink "face down" in order for them to drain and cool off a little. The drained lobsters can then be transferred to a large platter and brought to the table. It is a recommended practice to have an additional platter or two that will be used for depositing the shells and other unwanted body parts. Many people utilize the large pot that the lobsters were cooked in for this purpose, while other people often toss the remains into a **double-bagged** shopping bag. The sole purpose of using a pot, bags, or whatever else that works, is to reduce the number of items that will occupy the table. When several people are feasting on boiled lobsters, they will need room – and plenty of it.

If at all possible the proper tools should be available for each person. Many lobster pools and fish markets sell a "lobster eating kit" which includes all the essentials: a shell cracker, lobster pick, lobster fork, and a bib. Without the availability of these items, one may improvise by using the conventional nutcracker, ordinary kitchen shears, and a table fork. Given the necessary working tools and an abundant supply of paper towels, everybody should then be ready to lunge into the lobsters and begin *"The Feast!"*

Specific instructions for the eating of the lobster are as follows:

- Place the lobster on the plate and twist off the two large claws (this is assuming the lobster has two claws)
- Crack each claw with a shell cracker and use the lobster shears to help split the shell
- Remove the claw and knuckle meat
- Place the meat in a custard cup containing hot melted butter or margarine

- Separate the tail section from the body section by arching the tail section until the two sections separate
- Insert a lobster fork (the index finger actually works better) into the end of the tail section and push out the meat
- Use the lobster pick to mine out the tiny morsels of meat found in the body section of the lobster
- Break off the legs and try to push the meat out.

After The "Feast" Is Over

With the feast of lobsters being concluded – and hopefully enjoyed by all – it's time to clean up and dispose of the lobster remains. These two chores should be undertaken immediately since the lobster remains become very strong and objectionable with the passage of time.

It is recommended that the lobster remains be double-bagged and then be placed into a sturdy trash bag. The trash bag can then be taken to the dump or landfill. If such a facility is not available in the community, the bag may be deposited in one of the many dumpsters usually located in an around shopping centers. The dumpster should have a heavy steel lid because seagulls are known to make an awful mess of lobster remains. They are very adept at continuously pecking away at plastic bags until they have ripped them apart and yanked out the lobster remains.

The odor of the boiled lobster lingers for quite some time even after a good washing with soap and water. A swabbing of the hands and fingers with a lemon-

The Tale of the Lobster

scented towelette will quickly eliminate any objectionable odors. In the long run, the *"Feast"* should have been an enjoyable experience – and well worth all the effort!

Chapter 8. Some Commonly Asked Questions About Lobsters

Dear Reader: I have prepared for your indulgence 58 questions that I feel are pertinent to an understanding of some of the most important and interesting questions pertaining to the American lobster. These questions and their answers can also be used to impress your family and friends when dining on a feast of lobsters. Good luck!

Robert Delano Martin

1   Q   Where in the ocean does the American lobster live?

     A   In its natural habitat on the bottom of the ocean and in a range that extends from off the coast of Labrador in the north to off the coast of North Carolina in the south.

2   Q   Is there another name for the American lobster?

     A   Yes. It is also loosely referred to as a "Maine lobster," or a "Canadian lobster." Its counterpart in Europe, which is practically identical, is referred to as the "European lobster."

3   Q   Where do most lobsters "hide" on the bottom of the ocean?

     A   Under rocks, in rock crevices, under kelp, in mud burrows, and any debris that might litter the ocean bottom.

The Tale of the Lobster

4   **Q**   What species of marine life might endanger the lobster?

       A   Codfish, sandsharks (dogfish), cunners, eels, and generally any other type of groundfish.

5   **Q**   About how long does it take for a lobster to become a "chicken size" lobster?

       A   From 5-7 years in the wild depending upon seawater temperatures and other factors. In a controlled laboratory environment, it would take about two years to attain the same size.

6   **Q**   How does a lobster grow?

       A   Through a series of molts. When the meat mass inside the lobster outgrows its protective shell casing, the lobster must cast-off the outer shell. This process is known as "shedding" or "molting."

7   **Q**   Where are the most lobsters caught by lobstermen?

       A   Close to shore, called "inshore fishing."

8   **Q**   Where are most of the larger size lobsters caught?

       A   Usually in the offshore fishing grounds far out into the ocean and in very deep water depths.

9   Q   What organs does a lobster use to detect food, smell, and other situations required for living on the bottom?

    A   The lobster is endowed with thousands of sensory hairs and various appendages. The sensory hairs are called "satae" which function as organs of smell, taste, hearing, and possible endangerment. The lobster is also equipped with a pair of long branched-like antennae that serve as organs of touch.

10 Q   Are lobster fishermen the only people that catch lobsters?

    A   No. There are commercial dragger fishermen who bring up lobsters as part of their catch. There are also students who hold non-commercial licenses. There are also lobsters caught by longline fishermen. And lobsters are also pulled up in traps by "poachers." Finally, there are lobsters captured by SCUBA divers.

11 Q   Is a license required to fish for lobsters?

    A   Definitely! All different types of licenses are issued according to state laws.

12 Q   Could I just throw a few traps into the water to catch a few lobsters?

    A   Maybe! State laws apply and usually only a few number of traps would be allowed.

13 Q Would I get any resistance from local lobstermen by setting out a few traps?

A Maybe no, maybe yes. It all depends upon where you might put them. Any resistance would most likely occur in some Maine coastal waters where local lobstermen claim "territorial rights" and maintain vigilance over their fishing rounds.

14 Q What do you call a lobster with only one claw?

A A "cull" lobster. A lobster with no claws at all is also called a cull lobster that is usually boiled for its lobster meat contents by a local retailer of wholesaler.

15 Q What is the minimum size for a lobster to become a marketable lobster?

A In all lobster-producing states, it is 3¼ inches in carapace length.

16 Q What is the "carapace length?"

A The carapace length is a measurement of the lobster, taken from the rear of either eye socket to the juncture at where the shell of the carapace meets with the shell of the tail section (abdomen).

17 Q Is there also a maximum legal size?

   A Yes, especially in Maine and Massachusetts. It is a lobster more than 5 inches in carapace length measure, and has to be returned to the ocean.

18 Q What do they mean when they refer to a lobster as a "short"?

   A A short is a sub-legal size lobster that must be returned to the ocean.

19 Q Where does a female lobster store her eggs?

   A Upon extrusion and fertilization by the sperm of a male lobster (from some past mating experience), the eggs will become "cemented" onto the underside of the lobster's abdominal tail section. It will be several months before the eggs are hatched and released into the water column.

20 Q How many eggs will she hatch?

   A Thousands and thousands! The larger the female lobster, the higher the number of eggs that will be released into the water column. Oftentimes 50,000 eggs and more.

21 Q What is the "water column?"

   A It is commonly referred to as the area between the bottom of the ocean and the surface of the water above it.

22 Q What are the chances of survival of those thousands of eggs released into the water column?

A Not very good! It has been estimated that less than one-tenth of one percent (< 0.1%) of the larval lobsters dispersed into the water column and start their descent to the surface will survive several stages of molts and ever make it back down to the bottom of the ocean once again.

23 Q What was the weight of the largest lobster ever caught and reported on as such?

A About 45 pounds!

24 Q What is the purpose of the five blade-like appendages connected to the lobster's tail section?

A They are generally referred to as the "telson" and "flippers." They serve several functions, such as propelling the lobster backwards, and for burrowing out a mud tunnel. The underside of the telson also contains the terminus of the alimentary tract for disposal of solid waste material. Finally, one of the appendages of the telson is used for cutting a triangular "V-Notch" in a female lobster, if appropriate.

25   **Q**   What is the purpose of the "V-Notch?"

       A   It is a lobster preservation and conservation measure on the part of lobster fishermen in order to protect the egg-producing brood stock of the American lobster. Any female lobster brought to the surface with a "V-Notch" must be returned to the ocean. Also, any female lobster showing eggs or has a carapace length of more than 5 inches must be "V-Notched" and returned to the ocean.

26   **Q**   Are there any ways to distinguish a male lobster from a female lobster?

       A   There are several differences that are easy to identify. Please refer to Chapter 2.

27   **Q**   What is the ratio of male to female lobsters?

       A   Approximately 1:1, or 50% male and 50% female.

28   **Q**   Are all American lobsters the same color when purchased live?

       A   Usually. However, there are blue lobsters, red lobsters, and calico lobsters that are rarely seen at the consumer level.

29   **Q**   Where is the best place for buying live lobsters?

       A   At a lobster pool, fish market, a supermarket, or directly off the boat of a local lobster fisherman.

30 Q Where is the best place to buy cooked lobsters ready for eating?

A At the same outlets as for live lobsters (except from a local lobster fisherman).

31 Q Is there really any difference between a soft-shelled lobster and a hard-shelled lobster?

A Yes. A hard-shelled lobster will probably contain more solid meat, will have less body fluids and juices, and will be better eating. They are often called "winter lobsters." Soft-shelled lobsters are just the opposite and are often referred to as "summer lobsters."

32 Q Is the meat from larger size lobsters tougher and less tasty than is the case for smaller size lobsters?

A Not appreciably.

33 Q What is the most common size lobster sold at a lobster pool, fish market, or supermarket?

A About 1 to 1-1/8 pounds. They are commonly referred to as "chicken lobsters" or "chicks."

34 Q What times of the year are lobsters more expensive, less expensive?

A Lobsters are usually more expensive from late fall to late spring when the supply of lobsters declines, and less expensive during the summer and early fall months when the supply is greater.

35 Q What should I do if I buy live lobsters and find that one or more of them have "droopy claws" when I get them home?

A Eat them and enjoy them! Those lobsters are just a little sluggish, and they are perfectly fit to eat.

36 Q Will lobsters suffocate and die if I bring them home in a sealed bag or carton?

A No! Lobsters are capable of living for an extended period of time as long as they are not subjected to sudden and extreme changes in temperature.

37 Q Should live lobsters be placed in water when I get them home?

A Never. Fresh water can to lethally toxic to lobsters. So can household insecticides, copper, brass, and the like.

The Tale of the Lobster

38 Q   What should I do with live lobsters when I get them home and can't cook and eat them right away?

    A   They can be left in the same bag or carton and placed in the refrigerator.

39 Q   What is the safest way to handle a live lobster?

    A   By the shell of the carapace, which is the head and body (thorax) of the lobster.

40 Q   Some people say that lobsters "yell," "sing" or "cry" when they are plunged into boiling water. Is that true?

    A   No! That is an "old wives' tale."

41 Q   But don't they suffer a lot of pain?

    A   Not really. Death is virtually instantaneous.

42 Q   How long should a lobster be boiled?

    A   Most of the lobsters purchased by the retail trade are "chickens" that weigh about 1 to 1-1/8 pounds. They should be boiled for about 12 to 15 minutes. Larger size lobsters should be boiled for a little longer (Please refer to Chapter 7). However, it is important to note that the boiling time should begin <u>after the water has come to a second boil.</u>

43 Q What is the "tomalley" of the lobster?

A It is the liver of the lobster and is usually greenish.

44 Q What is the "coral" of the lobster?

A It is the undeveloped spawn in a female lobster, and is usually reddish.

45 Q What is the "blackish" material that is sometimes found in a boiled lobster?

A It is the ovaries of a female lobster that for some reason or another were absorbed. Just scrape it to one side and enjoy the lobster meat.

46 Q How many chicken size lobsters will I have to buy to get one pound of lobster meat?

A On the average it will take about 4-5 chicken size, two-claw lobsters to make one pound of clear lobster meat.

47 Q What is a "ghost trap?"

A A ghost trap is a lobster trap on the bottom of the ocean that has become detached from the buoy on the surface and the warp (rope) connected to the bridle of the trap. As such, the trap cannot be hauled to the surface.

The Tale of the Lobster

48 **Q** What causes a trap to become a ghost trap?

A By having the trap warp being intentionally cut, by being severed by a boat's propeller, as a result of commercial fishermen dragging the bottom, or by severe coastal storms.

49 **Q** What is an "escape vent?"

A Either a circular or rectangular opening in a lobster trap that allows sub-legal size lobsters to escape from the trap. It is a lobster conservation and preservation device required by law.

50 **Q** What is a "biodegradable panel" used for in a lobster trap?

A To allow lobsters of most any size to escape from a ghost trap stranded on the bottom. This is yet another lobster conservation and preservation measure required by law.

51 **Q** What does it mean to "scrub" a female lobster with eggs showing under the tail section?

A The molestation and the scrubbing of the eggs of a female lobster. A violation of state laws.

52 **Q** Do lobsters migrate?

A A few yes, but most do not. There have been instances of lobsters traveling considerable distances, but most lobsters do not have the inclination to relocate significant distances.

53 **Q** What is meant by the term "poaching" of lobsters?

A Poaching is the stealing of a lobsterman's lobsters or trap gear. It is the common practice on the part of some pleasure boaters, by some lobstermen themselves, and by anyone who has access to the shoreline. It is theft!

54 **Q** Do "offshore lobstermen" fishing in very deep ocean depths use the same type of trap gear as the "inshore lobstermen?"

A No. They utilize much larger and differently configured traps known as "bear traps."

55 **Q** Is the life of a lobsterman one that is comparatively easy and non-threatening?

A No. No. No. It might be seen as such by the summer tourist at dockside, but the lobsterman has to put up with a lot of bad weather at times (cold air, cold water spray, strong winds), and the like. Fishing for lobsters is hard and tedious work and can be extremely dangerous, according to the elements making up the foul weather conditions on the ocean.

56 **Q** Do most fishermen employ a "helper" or "sternman?"

A Yes. A helper or sternman (male or female) improves the lobsterman's efficiency and the amount of time he has to expend on hauling and setting his trap gear. In essence, it makes things easier for the lobsterman - and there is always another person on board if he gets into any sort of trouble while fishing his gear.

57 **Q** Where does the "spiny lobster" live?

A In the waters of the ocean south of North Carolina and throughout many parts of the world. The spiny lobster is characteristic of having no large claws, as does the American lobster, and is considered to be rather grotesque.

58 Q The American lobster has how many pairs of legs?

A Five pairs of legs in total, but the first pair of legs actually represent the "crusher claw" and the "ripper claw." They serve no real ambulatory functions at all. The four slender pair of legs serve many functions, the most important of which is that of enabling the lobster to move about on the ocean floor.

Chapter 9. A Dictionary of Important Lobster and Lobstering Terms

A

Abdomen

The abdomen of the lobster, which is located in the tail section.

Aft

The rear section of a boat. Also called the stern.

American lobster

Homarus americanus. Sometimes referred to as the Maine lobster. A marine invertebrate, a crustacean, a decapod, and an arthropod.

Anemone (Pronounced a-nem-o-nee)

A form of marine life that anchors itself to the bottom of the ocean and waiting to seize any unsuspecting prey.

Antennae

The pair of antennas that sweeps the water to detect the presence of pheromones, food, motion, and danger. In the American lobster the antennae are made up of the 1st antenna and the 2nd antenna.

Anterior

The forward end. The front end.

Anus

The posterior terminus of the lobster's intestinal vein. The organ used for ridding the lobster of solid waste material.

B

Back Shell

The hard shell of the carapace that encases the body section of the lobster.

Bait

The fleshy food that is used in lobster traps to entice lobsters into the traps. Usually some sort of fish offal such as herring, redfish, and flounder cuttings, pogies, and the like. Sometimes referred to as "pickle," "brim," and "gurry."

Bait Apron

Usually a heavy vulcanized apron worn by the lobstermen to fend off bait and other matter aboard a lobster boat.

Bait Barrel

A large wooden barrel used for the storage of bait aboard a lobster boat.

Bait Iron

A device like an ice pick for holding a rack of fish to be transferred onto the bait spindle in a lobster trap. Also called a "spudge iron," "spudgeon," "bait stick" and "bait needle."

Bait Shack

Usually a wooden structure on a dock or pier that is primarily used for the storage of lobster bait.

Ballast

Weight added to lobster traps to make them sink to the bottom and to make them set well on the bottom. Bricks are the most common forms of ballast.

Barrel

One of the terms used by a lobster fisherman and others in the trade to denote a lobster with both claws missing.

Barrel Hoist

A pulley and wire cable that is primarily used to raise and lower bait barrels in and out of the lobsterman's boat. Especially helpful during low tides. Are also used to raise lobsters and other gear up onto a pier or dock.

Berry

The eggs of a female lobster. A female lobster is said to be "in berry" when ova is showing on the underside of the tail section.

Bobber

See Toggle Buoy.

Body

The head and thorax section of the lobster that is covered by the hard shell of the carapace. The cephalo-thorax.

Bow

The front and forward end of a boat.

Breaking Plane

A specific point on the ischium appendage that separates when a lobster "throws" or "shoots" one of its large claws.

Bridle Warp

The rope network that is tied to the lobster trap for securing the main warp leading up to the main buoy on the surface.

Brim

Lobster bait, specifically ocean perch. See bait.

Bug Chasers

A slang term used by some lobstermen when referring to lobster marine scientists.

Bullet

A term used to denote a lobster with both claws missing.

C

Cannibalism

The tendency of lobsters to prey on others of the same specie, either dead or alive.

Canyons

Referring to the 14 canyons of the Continental Shelf that extend along a certain stretch of ocean off the eastern seaboard of the United States.

Car

See Live Car.

Carapace

The hard shell of the lobster that encases the body section. The part of the lobster that is measured to determine if it is of legal size.

Carpus

One of the knuckle appendages between the large claw and the body section of the lobster.

Catch

The number of pounds of legal size lobsters landed.

Chafing Gear

A wood or metal reinforcement over a small section of the hull on one side of the lobster boat. Protects the hull of the boat from being chafed or damaged.

Chelipeds

The large "crusher" and "ripper" claws of the lobster.

Chitin

The composition of shell structure of the lobster.

Cinch-Up Tag

A "back-up" tagging device used in research studies conducted by lobster marine scientists. Usually is snugly attached around the carpus appendage of the ripper claw or around the ripper claw itself.

Cockpit

The open area from the wheelhouse to the stern (rear) of a boat.

Coolie

A slang term for a lobsterman's helper or sternman.

Coral

The undeveloped spawn in a female lobster.

Counter

A legal size lobster.

Crusher Claw

The usually larger and bulkier of the two large claws of the lobster. Used to seize and crush food, shellfish, and other marine life. Also used for defense and attacking prey.

Cull

A lobster with only one large claw. Also a lobster without either a crusher claw or a ripper claw.

CL

An abbreviation for Carapace Length. See Carapace.

D

Dactylus

The movable part of the lobster's large claws. Also found on the first two pairs of the lobster's slender walking legs.

Depth Finder

See Fathometer

Double Gauge

A brass or bronze gauge that is used to measure a lobster's carapace to determine legal size

Doubles

A string or trawl of lobster traps consisting of just two traps.

Down East

A questionable geographical area along the Maine coast starting beyond Ellsworth and proceeding along the coast in a northeasterly direction.

Drag

A rather crude wooden device used by lobster pound owners to drag out lobsters on the bottom of a lobster pond.

Dragger

A commercial fishing boat that drags its heavy nets along the bottom to catch groundfish and shell fish.

E

Effort

The number of lobster traps used in the catching of lobsters

Egger

See Berry.

Endopodite

One of the blade-like appendages of the lobster's tailfan.

Escape Vent

One or more openings in the lobster trap that is designed to allow for the escape of sub-legal size lobsters and other small species of marine life.

F

Fan Tail

A wooden or vinyl spindle protruding from the top of a lobster buoy.

Fathom

A nautical unit of measurement. 1 fathom = 6 linear feet.

Fecundity

The egg-bearing capacity of a female lobster.

Feelers

The two long sweeping antennae of the lobster. Specifically, the 2nd antenna and the flagellum. Used for sensing food, chemicals, food, and other objects in the water.

Fish Rack

A spindle in the "kitchen section" of a lobster trap that is used for storing bait. Also the bait itself.

Fishing Head

The meshed netting part of a lobster trap that lobsters and crabs crawl up and into in order the get at the bait in the "kitchen section" of the trap.

Flatfish

Types of groundfish such as flounder and sole.

G

Gaff

A wooden pole with a steel hook device at one end. Used to gaff the lobster buoy and warp out of the water.

Gaffkymia

A usually fatal disease in lobsters. Also known as the "red tail" disease. Usually brought on by a shell disease while lobsters are being held in storage.

Gas Disease

A usually fatal disease in lobsters. Mainly attributed to an insufficient supply of oxygen in the saltwater while lobsters are being held in storage.

Gauge

A brass or bronze gauge used to determine if a lobster is of legal size. See Double Gauge.

Gear

Any equipment or material that is used in the conduct of lobster fishing. Examples are traps, warp (rope) buoys, barrels, etc.

Gear Conflict

A conflict between the traps of lobster fishermen and the nets of dragger fishermen.

Ghost Trap

A lobster trap (or pot) that has become separated from the main warp leading to the buoy on the surface and making it impossible for the trap to be hauled to the surface.

Gonopods

The first pair of swimmerets (pleopods) in the male lobster. Also referred to as "stylets."

Great Chelipeds

The large and predominant claws of the American lobster.

Green Glands

Organs of digestion in the lobster that are used to excrete fluid waste material.

Groundfish

Fish that inhabit the bottom of the ocean, such as codfish, haddock, cusk, flounder, eels, sand sharks, and the like. Usually caught by draggermen and longline fishermen.

Gunwale

A horizontal surface on either side of a boat where the lobster fisherman hauls his traps up and onto. Also referred to as "washrail" and "washboard."

Gurry

See Bait.

H

Head

The circular meshed entryway into a lobster trap.

Helper

A person who is a lobsterman's helper and who usually works in the open cockpit of a lobster boat. More frequently called a "sternman." Can be either male or female gender.

Herring Bag

A nylon or twine meshed bag used for holding lobster bait, such as the skeleton remains of herring and flounder.

Homarus americanus

See American Lobster.

Hull

The framework of a boat, especially the bottom and sides.

I

Inshore Lobstering

The fishing for lobsters relatively close to shore and in much more shallow water than is the case for offshore lobstering.

Interlock Mechanism

A mechanism in one of the knuckle appendages of the large claws that prevents the claws from being twisted or ripped off.

Invertebrate

An animal without a backbone or internal skeleton.

J

Joints

Movable appendages in the lobster such as the knuckles, walking legs.

Jumbos

Larger size lobsters usually weighing three pounds or more.

K

Keeper

See Counter.

Kelp

A form of seaweed. A popular hiding place for lobsters. Also a source of food for lobsters.

Kitchen

The section of the lobster trap that is entered by the lobster to get at the bait stored there.

Knuckles

See Joints.

L

Larvae

Larval lobsters floating and drifting on or near the surface of the water. Also lobsters swirling around in circular rearing tanks at a lobster hatchery. Referred to by lobster marine scientists as "fry."

Live Car

See Lobster Car.

Lobsterling

A fifth or six stage larval lobster that goes to the bottom to stay and live.

Lobster Car

A floating crate-like device used by the lobsterman to temporarily store his catch. Also called a "car" and "live car."

Lobster Tank

A tank for temporarily holding lobsters. Used on most lobster boats and in lobster pools, fish markets, supermarkets, and some seafood restaurants. Also called a "holding tank."

M

Mainliner

A term denoting a very successful fisherman.

O

Offshore Lobstering

The fishing for lobster far offshore and in much deeper water.

Outsider

Any person fishing for lobsters who is not a resident or native of the town. Badly looked upon by full-time lobster fishermen, especially in "Down East" Maine coastal waters.

Ova

An egg mass.

Oviduct Openings

The two openings in the female lobster through which the ova pass from the ovary to the outside.

Oviferous

Carrying ova.

Oviparous

Producing eggs that hatch after leaving the body.

Oviposition

The process of laying eggs in the female lobster.

P

Palinurus

The spiny lobster that inhabits the much warmer water south of North Carolina and throughout many bodies of water throughout the world. A "lobster" without two large claws and caught mostly for the meat contents of the tail section.

Paper Shell

See Shedder.

Parlor

One or more sections of a lobster trap that the lobster enters after eating the bait in the "kitchen" section of the trap.

Parlor Head

The hooped opening of the lobster trap that the lobster crawls through to get out of the "kitchen" and into the "parlor" section of the trap.

Part-Timer

A person who fishes for lobsters on a part-time basis, such as a student.

Peg

A pointed wood or plastic plug used to keep the large claws of the lobster shut-tight. Not in common usage anymore. See Band.

Peg Box

A small wooden box used for storing rubber bands and pegs. Sometimes used by the lobster fisherman to set a lobster into for banding or pegging.

Pheromone

Any of various chemical substances secreted externally by the lobster to convey and produce specific responses to other lobsters.

Pilot House

See Wheel House.

Pleopods

See Swimmerets.

Plug

See Peg.

Plug Rot

A blackish color around the lobster's large claws that are plugged shut with a wooden peg. With the predominant use of rubber bands, plug rot is no longer a major problem.

Poaching

The act of stealing lobsters from the traps of a lobsterman. A very serious practice of theft that is usually carried on under the cover of darkness. Lobstermen have their own subtle but effective ways to deal with poachers.

Pool

A lobster dealer's usual place of doing business. Generally found at a lobster pool or fish market.

Posterior End

The backward end of the lobster.

Pot

See Trap.

Pound

A sealed-off cove or inlet where seawater flows in and out, but where impounded lobsters cannot escape. Used to hold large quantities of lobsters for use during the colder winter months when they are scarcer.

R

Rattlers

See Shorts.

Recruit

A sub-legal size lobster that has now grown in size following a molt and then becomes of sufficient size to be measured as a legal size lobster, presently 3¼ inch CL measure. See Carapace.

Redfish

A popular type of lobster bait. Its biological name is Ocean Perch.

Red Tail

See Gaffkymia.

Regeneration

In the lobster, the process of restoring a lost appendage. Also the process of forming of a new shell as a prelude to a molt.

Ripper Claw

The usually smaller and more slender of the large claws of a lobster. Sometimes referred to as the "pincer claw" and the "pincher claw."

Rock Lobster Tag

A tag device used in lobster research studies. The tag is anchored at the juncture of the carapace and tail section.

Roe

The swollen ovaries or expelled eggs of certain crustaceans.

Rope

Used to haul traps to the surface. More commonly called "warp."

Rostrum

The sharp and pointed structure that juts out from above and between the eyes of a lobster.

S

Sand Flea

A marine organism that devours the bait in a lobsterman's traps.

Satae

The sensory hairs of the lobster.

Scrubbing

The molestation and removal (by scrubbing) of a female lobster's eggs carried on the underside of the tail section.

Sculpin

A groundfish known for eating the lobsterman's bait. Perceived as being rather "ugly."

Scupper

The openings cut into the waterway of a boat so that the water on the deck floor can drain out into the ocean.

Sea Urchin

A form of marine life that features sharp and spiny-like needles. In the jargon of some lobstermen, they are referred to as "whore's eggs." While they are despised by many lobstermen, the sea urchin is retained by some and sold for its meaty flesh.

Seeders

Female lobsters that have become oviferous (egg-bearing), usually while being held in storage in a lobster pound.

Seeder Program

The practice of a state agency to purchase female lobsters from pound owners because they have become oviferous.

Set-Over Day

A day that a lobster trap is not hauled by the owner of that particular lobster trap.

Shed

See Molt.

Shedder

A lobster that has recently molted.

Shell Disease

See Gaffkymia.

Ship-Shape

A boat that is cleaned up and in order again.

Shoot A Claw

See Throw A Claw.

Shorts

Sub-legal size lobsters. Illegal to retain. Also called "snappers."

Singles

The fishing of single traps. See String.

Skiff

A small boat used by the lobsterman to go back and forth to his lobster boat tied up to a mooring.

Skipper

The captain of a boat.

Sleeper

A lobster with droopy claws.

Smack

An earlier type of small fishing boat and usually equipped with a sail.

Snappers

See Shorts.

Sole

A type of groundfish similar to the flounder, but having more value.

Somites

The six accordion-like sections of the lobster's tail section (abdomen).

Spaghetti Tag

See Sphyrion Tag.

Spawn

The developing egg mass within the ovaries of a female lobster.

Sphyrion Tag

A tag device used in lobster research studies. The tag is anchored at the juncture of the carapace and the tail section.

Spiny Lobster

See Palinurus.

Spudge Iron

See Bait Iron.

Spudgeon

See Bait Iron.

Statocysts

A small organ in certain of the lobster's walking legs that contains sensory hairs and sand granules that are used for maintaining balance when the lobster is walking along the bottom of the ocean.

Stern

The rear of a boat. Also referred to as "aft."

Sternman

See Helper.

String

A series of more than two lobster traps connected one to another and usually with a marker buoy at each end of the string. Also called a "trawl."

Stylets

See Gonopods.

Summer Lobsters

Lobsters that are soft-shelled following a molt. See Winter Lobsters.

Swells

The rising and falling of the ocean water on the surface.

Swimmerets

The several pairs of paddle-like appendages on the underside of the lobster's tail section (abdomen). Also referred to as "pleopods." Serve many functions, and more importantly as a depository after a female lobster has extruded her eggs from the oviducts.

T

Tail

The tail section of the lobster. More properly, the lobster's abdomen.

Tailfan

The most posterior section of the lobster and containing five flat blade-like appendages.

Telson

The middle appendage of the lobster's tailfan.

Thick-A-Fog

A Maine lobsterman's expression when there is a "pea soup" fog on the ocean. "It's thick-a-fog out there!"

Throw A Claw

The act of a lobster giving up a large claw in order to escape from an enemy and a worse fate. Also called "shooting –a-claw," "reflex amputation," "self-amputation," "autotomy," and "defensive mutilation."

Toggle Buoy

A device that is tied to the main warp (rope) leading to the trap. Usually consists of two round pieces of cork that help to keep the warp off the bottom at low tide or in shallow water. Also called a "bobber."

Tomalley

The organ of digestion in the lobster. Characterized by its softness and green color.

Trap

A wood or wire device set on the bottom of the ocean to catch lobsters and crabs. The poly-vinyl-clad wire traps are being used extensively by most lobster fishermen. Also called a "pot."

Trawl

See String.

Triplets

Fishing three traps in a string and tied one to another.

V

V-Notch

A 'V"- notch cut into one of the flat blade-like appendages making up the lobster's tailfan. Instituted by Maine lobstermen to protect the present and future brood stock of the fishery.

W

Walking Legs

The four pair of slender thoracic walking legs of the lobster.

Warp

A commonly used nautical term for rope.

Washrail

See Gunwale.

Water Column

The water between the bottom of the ocean and the surface of the water above it.

Weather

A nautical term that often implies bad weather. "There's a lot of weather out there!"

Wheel House

The forward deckhouse of a boat that contains the wheel, navigational equipment, hydraulic trap-hauler, and the like. The wheelhouse is where the lobsterman stands when he is fishing for lobsters. Also called a Pilot House.

Winter Lobsters

Hard-shelled lobsters. See Summer Lobsters

Worm Borers

A marine specie that bores through the lathing (wood). Applies only to wooden traps.

Y

Yield

See Catch.

Chapter 10. Let's Go Fishing - A Lobstering Pictorial

- ➢ **On The Ocean With A Maine Lobsterman**
- ➢ **The Gear**
- ➢ **The Lobsters**

In Chapter 5, "The Lobster Fisherman," an emphasis was placed on the lobsterman, his boat, his gear, his fishing, and his lifestyle. In that chapter there was the occasion of being aboard the lobster boat and experiencing a day on the ocean while fishing for lobsters with a Maine lobsterman. This, the concluding chapter, provides the reader with a pictorial account of some of these happenings, as seen through the eyes of the photographer, and captured through the lens of the camera.

Robert Delano Martin

Serenity! An early morning calm prevails inside the beautiful harbor at Cape Porpoise, Maine. Jonesport and Novie lobster boats are at their moorings. The shack-like object in the foreground is a floating "lobster car" where one or more lobster fishermen store their lobsters and lobstering gear.

After loading up with the bait of the day, a lobsterman does a U-turn before heading out of the harbor at Cape Porpoise, Maine. The small boats in the foreground are used by the lobstermen to get to and from their boats tied up at their moorings in the harbor.

Robert Delano Martin

Lobsterman Arnold "Joe" Nickerson III pilots his boat to dockside to load on bait and pick up his helper. Like a good number of lobster fishermen, Joe Nickerson fishes from the starboard (right) side of the boat. The names painted on the bow of the boat depict the names of his two daughters, Kori and Anne.

The Tale of the Lobster

What shall we serve the lobsters for dinner during the next day or two? It shall be the herring in the bait barrels that lobsterman Joe Nickerson is lowering into his boat. The "NICK" painted on the bait barrels denotes that the barrels are the property of Joe Nickerson. All of his gear is identified by either his name or his license number, in his case "3875."

Robert Delano Martin

With one hand on the throttle and the other hand on the wheel, lobsterman Joe Nickerson heads out of Cape Porpoise Harbor and toward the first trap to be hauled. Because most lobstermen in the area fish relatively close to shore (Inshore lobstering), the first trap to be hauled is generally no more than ten minutes away.

The Tale of the Lobster

While the lobsterman's helper is kept busy forking bait, Joe Nickerson hauls his first trap of the day. The hydraulic trap hauler is controlled by a brass lever under his left hand. The trap hauler is located at just about knee-level and directly in front of where the lobsterman is standing.

Lobsterman Joe Nickerson grasps a hold of the trap's bridle when it reaches the side of the boat. This is the "wettest" part of the job for the lobster fisherman. Because of the various forms of marine life that is hauled up in most traps, this is often the "messiest" part of the lobsterman's job.

The Tale of the Lobster

Up onto the gunwale comes the lobster trap. The trap warp (rope) that has been piling up is booted aside, out of the way, just behind the lobsterman's left boot This particular trap features three individual compartments, made up of a "kitchen" compartment where lobsters first enter the trap, and two "parlor" compartments where lobsters enter when trying to find a means of escape. The shiny rectangular-shaped devices near the end of the trap are two "escape vents" that allow sub-legal size lobsters to crawl through to in order to escape from the trap.

The lobsterman and his helper are kept busy cleaning out the contents of the trap. In order to work efficiently, each must work in concert with the other. While the lobsterman removes the contents of the traps, the helper is kept busy removing the old bait and replacing it with fresh bait. For the most part, this activity will be repeated time after time until the last trap haul of the day has been accomplished.

A "keeper" or not a "keeper?" The lobsterman puts the brass gauge to a lobster removed from one of his traps. If the lobster measures out to be of legal size, then that lobster will be retained as a marketable lobster. If it does not satisfy the legal size requirements, then the lobster must be returned to the ocean once again. The minimum legal size is 3¼" (carapace length measure) in all lobster-producing states.

Robert Delano Martin

A lobster fisherman who hauls and resets his traps close to shore is rarely alone. Other lobster boats are usually well in view, as well as the scraggy shoreline shown in this photograph. It is a common occurrence for one lobsterman to pull up alongside another and have a little chit-chat. And what do they talk about? Usually the weather, the catch, and other "lobster talk" in general.

The banding of a lobster's large crusher and ripper claws usually befalls upon the lobsterman's helper. The task is performed as soon as possible in order to prevent any injuries to other lobsters being kept in a temporary holding tank. The banding of the large claws is facilitated through the use of a special tool that spreads open the rubber band so that it may be slipped over the claw, thereby disabling it

Where will I reset my traps? In this photo, lobsterman Joe Nickerson is looking for a rocky substrate on the bottom of the ocean. When he sees such a rocky substrate on his fathometer/depthfinder, the traps will be pushed off the boat, one at a time, and back into the ocean once again. The lobster fisherman must be quite careful during the resetting of his trap gear to avoid them descending upon the gear of other lobstermen fishing in the same area.

The Tale of the Lobster

Baiting-up is not one of the most envious chores that has to be carried out aboard a lobster boat. It is, however, one of the most important tasks to be accomplished and is generally the responsibility of the lobsterman's helper. In this photograph, a nylon mesh bait bag is being stuffed full with the bait of the day - large herring,

What a mess! Tangled warp and buoys are a common occurrence when there are many lobstermen fishing their traps within a small area. By closely scrutinizing this photograph, one can see lobstermen fishing their traps within a small area. By closely scrutinizing this photograph, one can see fellow lobsterman as well. Notice that the color and shape of the buoys are not alike. It is this type of occurrence that interferes with the efficiency of work on a lobster boat, and if the catch is off and the weather is foul, it makes it even a tougher day for both the lobster fisherman and his helper.

The Tale of the Lobster

Repairs to lobster gear are usually done right on the spot unless the damage is so extensive that it is better undertaken on shore. In this photograph, lobsterman Joe Nickerson is in the process of tying new nylon mesh twine into the netting leading from the first parlor compartment to the second parlor compartment.

Robert Delano Martin

When a lobster fisherman feels that his traps are not yielding a satisfactory catch, he will most likely move the gear to another location. The traps will be stacked atop the gunwale or aft on the rear deck. At the call of the captain, the helper will shove the trap off the boat and into the ocean. As the trap warp begins to slacken and begins to disappear from sight, the buoy will then be thrown in after it.

The Tale of the Lobster

Most lobstermen carry a lunch with them, and that lunch is usually eaten much earlier than one might expect. In this case, the time for lunch was just about 10 AM. That is probably not too early when one takes into account that this lobsterman was up and out of bed some five to six hours earlier, or more. It will probably be the only time that the lobsterman and his helper will not be hard at work hauling and setting his traps.

Robert Delano Martin

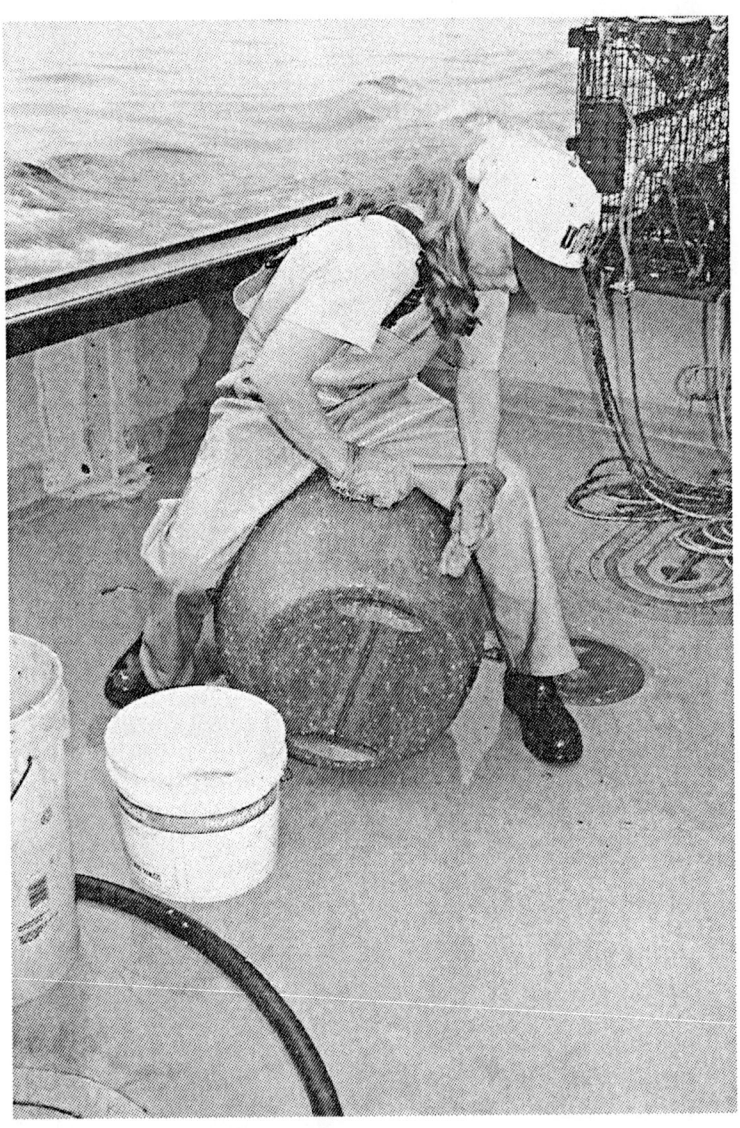

"Scrubbing-Down" is yet another chore to be performed by the lobsterman's helper. It is usually carried out when returning to port after the last trap haul of the day has been made. Bait barrels, buckets, and the interior of the boat are all washed up and scrubbed down. A high-pressure stream of fresh seawater is fed through a hose to wash down the entire boat from bow (front) to stern (rear). Most lobster fishermen are very particular about the cleanliness of their boats. They don't want to fish from a stinky, smelly, and dirty boat! They also realize that an unclean boat is an attraction for seagulls that often have the nasty habit of leaving their "calling cards!"

The Tale of the Lobster

Free lunch! Any edible residue from a day of lobstering ends up in the ocean and brings about the attraction of seagulls. A "squabble" of seagulls will stay aft of a lobster boat for long periods of time in the hope of scooping a morsel or two of food from the ocean.

The "catch of the day" is usually sorted out (also called "culled out") by size, weight, condition, and so forth. Hard-shell lobsters are ordinarily separated from their soft-shell counterparts, and two-claw lobsters are kept separate from cull lobsters (either one or both claws are missing) because the two-claw lobsters are more readily marketable and yield a somewhat higher price per pound. This activity is ordinarily done on the lobsterman's boat or at his or hers wholesaler's "live car" floating in the harbor.

The Tale of the Lobster

A moment of truth! Lobsterman Joe Nickerson has a trusting relationship with his lobster dealer. After weighing the lobsters, they will be placed into wooden crates and suspended inside or outside of a floating "lobster car." The lobsters will stay there for an indefinite period of time and until such a time that the lobsters will bring the best price possible in the marketplace.

One of the final tasks aboard the lobster boat will be that of hoisting the empty bait barrels up and onto the dock. The same barrels will be used for bait for the next day out fishing. If fuel has to be taken on, that is accomplished by most lobstermen at day's end. After these chores have been taken care of, Joe Nickerson will pilot his boat out into the harbor and secure it to his mooring. Once back on shore, a hard day's work will be over, and he will then have time to pursue something else - other than lobsters!

The Tale of the Lobster

The Lobster Buoy - State laws requires that each licensed lobster fisherman display his "colors" on either side of the bow or at an elevated position atop the lobster boat. Most lobstermen, like Joe Nickerson of Cape Porpoise, Maine, elect to choose the latter method. In this photograph, Nickerson's red and yellow buoy is attached to the antenna mast associated with his marine radio.

Robert Delano Martin

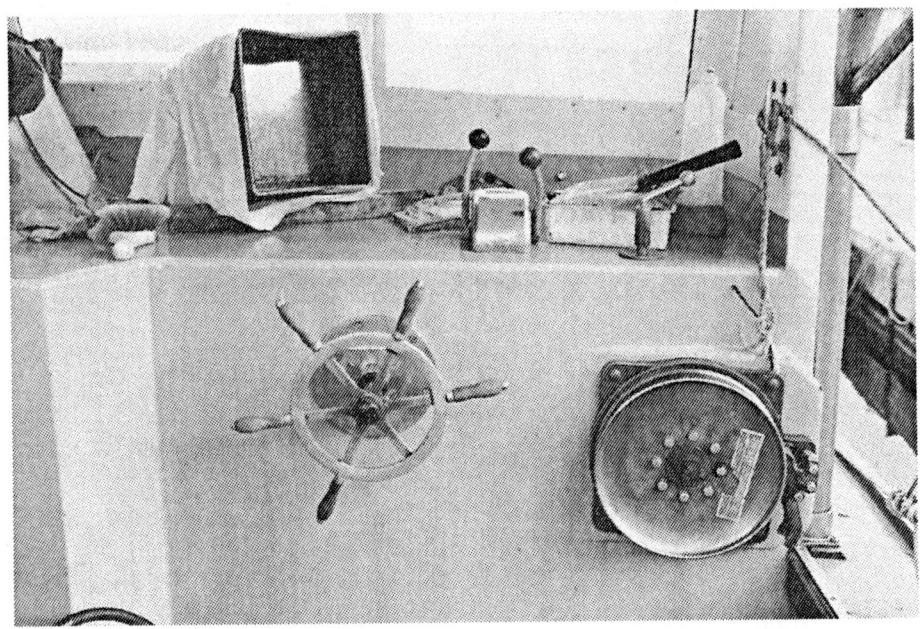

Some of the "working tools" aboard a lobster boat can be observed in this photograph of the wheelhouse. They include the following: the fathometer/depthfinder, which measures the depth of water and indicates the type of substrate on the ocean bottom. There is also one lever that controls forward, neutral, and reverse speeds; beside that lever is another lever that controls the action of the hydraulic trap-hauler, and the round-shaped trap-hauler itself (below right). There is a hammer that is "kept at the ready" which is used for making any on-the-spot repairs to lobster gear and for inflicting a stunning blow to unwanted and pesky groundfish that manage to get into traps and steal the lobsterman's bait (eels, sculpins, dogfish, and the like). The pilot wheel is located pretty near dead center in the wheelhouse. Not shown in this photograph is the marine radio that a lobsterman depends upon to chat, listen to weather forecasts, and for use in a time of emergency. Also not shown is the Loran tracking gear that is used for fixing positions on the ocean.

The Tale of the Lobster

The Lobster Trap - Many lobstermen fish all three-compartment traps that consist of a "kitchen" compartment and two "parlor" compartments. Escape vents are inserted in one or both of the parlor compartments to allow for sub-legal size lobsters to escape from the trap, but restricts most legal size lobsters from doing so. Vinyl-clad wire traps have pretty much replaced the older conventional wooden traps because they are lighter, hold the ocean bottom better, and stack better when moved on and off the lobster boat. It is the opinion of some, if not most, lobstermen that the wire traps fish better and catch more lobsters.

The Toggle Buoy - Most lobster fishermen, when fishing their traps in waters where there are extreme changes in water depths and when there are strong tidal currents, will employ the use of a toggle buoy to keep the warp off the bottom. The purpose of this is to keep the warp from becoming entangled with trap gear, rocks, and any other artifacts that may be on the bottom. The toggle buoy is usually made of cork or Styrofoam and is tied to the main warp a few feet from the surface.

The Bait - The lobster fisherman might use several types of bait throughout the year because he often has to accept from the "bait man" only that type of bait which is available to him. The baitfish shown in this photograph is herring, which is a favorite of many lobster hunters. When the skeletal remains of other fish are used, such as red fish (Ocean Perch) and flatfish, they also must be stuffed into nylon mesh bait bags, as depicted in the above photograph.

The Double Gauge - This is truly a unique tool for measuring a lobster to determine if it is of legal size. While all lobster-producing states adhere to the minimum legal size of 3¼" (carapace length measure), Maine also has a "double gauge law," whereby a lobster that measures more than 5" (carapace length measure) must be returned to the ocean. It is a strong contention of many of Maine's lobster hunters that these larger lobsters represent the present and future brood stock of the Maine lobster fishery. So it can be said that Maine has both a minimum size gauge and a maximum size gauge. Other restrictions on lobsters, other than size, also have an impact on what lobsters can be retained and what lobsters must be returned to the ocean. These restrictions are discussed throughout this writing.

The Tale of the Lobster

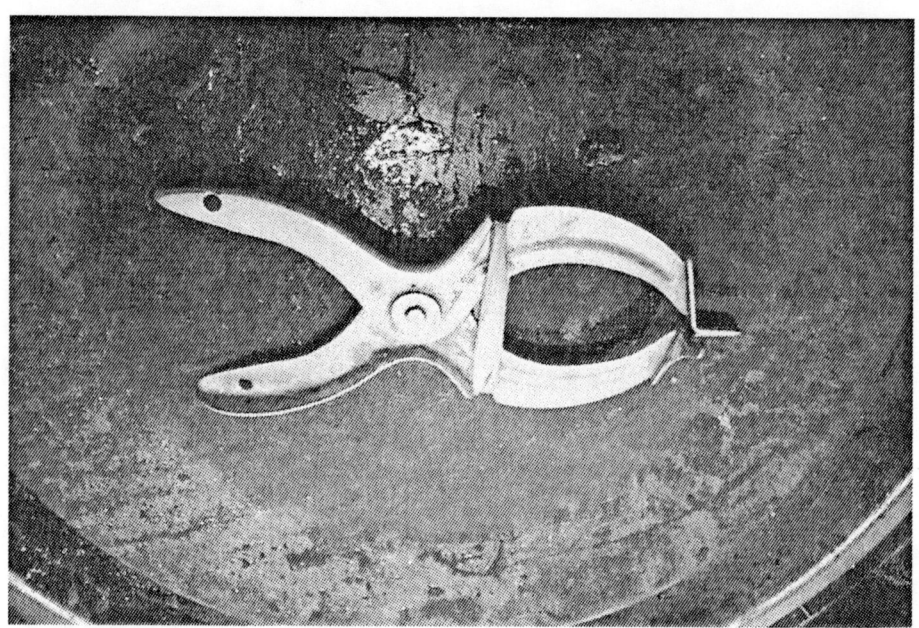

The Rubber Band Spreader - This ingenious device is used to spread a very small but very strong rubber band around both the crusher claw and the ripper claw of a lobster. In principle, the spreader works in a manner exactly opposite to that of a pair of ordinary household scissors. A rubber band is slipped over the end of the spreader, the handles are squeezed together to spread open the band, and the band is slipped over the large claw of the lobster. Then, with a flip of the wrist, the band is released from around the claw and the claw becomes immobilized. The banding of the large claws of lobsters is necessitated because of the cannibalistic tendencies of lobsters being held in captivity. The use of rubber bands is much preferred to the older practice of "plugging" the claws shut, because a wooden or plastic plug injures the flesh and can lead to disease. A plug can also cause a blackish color to form, called "plug rot," which is distasteful to a person who is going to eat that lobster. The plugging of lobster claws is virtually a practice of the yesteryear.

Robert Delano Martin

A perfect lobster in every way! This lobster is a splendid specimen because all of its appendages and body parts are intact. There are no deformities and the lobster's many parts are of the proper shape and form. However, this particular specimen portrays a "left-handed lobster" because the larger and more bulky crusher claw is on the left and the longer and more slender ripper claw is on the right.

The Tale of the Lobster

This photograph illustrates both the strength and tenacity of the lobster's claw, in this case the "ripper claw." This particular lobster would not yield the photographer's pen and the claw had to be physically opened by hand in order to retrieve the pen. This is yet another good example of why lobstermen band-shut both of the powerful, large claws of lobsters.

This is a beautiful specimen of the American lobster - but it could not be retained by the lobsterman. Two reasons: First, this female lobster was as "egger" (bearing eggs on the underside of the tail section). Secondly, she measured more than 5" in carapace length. This lobster was returned to the ocean.

The Tale of the Lobster

This lobster is classified as a "cull lobster." As such, it will command a lower price than would be the case of a lobster having two well-formed claws. The small "ripper claw" suggests that one of two events could have taken place: the abnormal small claw might have been the result of an unsatisfactory molt, or it is still in the process of regeneration after having lost its original claw in mortal combat with another combatant on the ocean floor.

The is another specimen of a "cull lobster." It will be noticed that the long and sleek "ripper claw" is intact, but the more bulky and dominant "crusher claw" is missing. This photograph reveals a very important distinction between the two claws. It can be seen, looking very closely, that the ripper claw features numerous sharp and pointed teeth. The crusher claw, on the other hand, possesses no sharp and pointed teeth (not shown), but does feature almost flat, molar-like structures. It is conjectured that the missing crusher claw was severed by a predator or voluntarily "shot off" by the lobster in order to escape a worse fate in its attempt to escape from a predator.

The Tale of the Lobster

This "berried" female lobster carries on its underside tens of thousands of small eggs. Because these eggs are dark brown in color, this lobster will have to carry and protect them for ten or eleven months before they will be ready to be hatched out into the water column. The eggs are fastened to the swimmerets and underside of the tail section by a glue-like substance. When the fosterage period has come to an end, the lobster will shake off her eggs, and over perhaps a few days all of the eggs will float to the surface to begin their dangerous and perilous life near or on the surface of the ocean.

Robert Delano Martin

Epilogue: Be Merciful Unto The Lobster

Homarus americanus, known as the American lobster, and even the so-called Maine lobster, faces an uphill battle for survival from its very moment of conception. In the 1600's the American lobster was judged so plentiful and so easily caught by hand that it was looked upon as being of little intrinsic value. As abundant as the lobster was, it was often used as a fertilizer rather than being served as a delicacy from the ocean's depth. There have been times in later years that the American lobster has been so extensively and intensively fished that it has been referred to by some marine scientists as being an endangered specie. Indeed, the life cycle of the lobster, from its birth to it's death, is one of total endangerment!

As we have seen, the life cycle of the lobster starts only after the female lobster has molted, is in a soft-shell condition, and seeks out a male lobster to mate with. Ten or eleven months will pass before the female lobster's eggs are ready to be extruded from the body. The eggs, in the thousands, will flow in a steady procession along the underside of the body to meet up with the sperm of a male lobster from a previous mating experience. The profusion of eggs are then "cemented" as a mass onto the underside of her abdominal tail section. For another ten to eleven months the female lobster will carefully nourish, clean, and protect her brood from other marine species that live on the ocean bottom. When the time is biologically correct, she will release her young into the water column where they will begin their ascent to the surface. And here is when and where the carnage begins to reduce the number of larval lobsters that will actually make it to the surface. Those that do make the journey and reach the surface now become easy prey to the perils of the ocean.

There are fish, ocean birds, and the possibility of being washed ashore upon rocky coastlines, sandy beaches, and outcroppings of ledge. Five or six stages of larval development will take place on or near the surface before the final larval stage is reached and when the tiny lobster will be able to descend to the bottom of the ocean where it will seek out a life of seclusion and hiding.

The odds of a larval lobster ever making it to the bottom again are not in its favor. One lobster marine scientist in particular has estimated that **less than one-tenth of one percent (< 0.1%) of the lobsters hatched out on the ocean floor will ever return there again!** This is an extremely high mortality rate.

Upon descending to the bottom, the tiny *"lobsterling"* will immediately seek out a life of hiding and seclusion. This will be a very hostile environment for that lobsterling. Venturing out, if at all and only to seek food, the new inhabitant of the deep will hide under rocks and any other substrates in the immediate area. Not much is seen of the lobsterling until it grows to the extent that it is somewhat capable of fending for itself in the open. It will then start to move about in search for food, usually plankton, and even the bait that is stored in the lobsterman's traps. And here is when the chaotic life of the lobster on the bottom really begins.

Over a period of several years, that lobster has probably been hauled up in the lobsterman's traps hundreds of times and has occasionally received some pretty tough handling in the process. Just imagine the trauma that the lobster must experience while being raised to the surface in its prison, the lobsterman's trap. Time after time and year after year that lobster will be hauled to the surface, until one day a lobsterman will put his gauge to it and lament, "You're a *'keeper'* now. I've finally gotcha!"

In the lobster fishery there are lobsters that are referred to as *"culls."* Cull lobsters have most likely come by that name as a result of being *"culled out"* of the lobsterman's catch because one of the large claws is missing and oftentimes are of less value. And as was discussed in Chapter 8, a *"cull"* can also be a lobster with both of its large claws missing.

"Cull lobsters" have gotten that way because they lost one or both of their claws in a battle with an adversary or as a direct result of being mishandled by people in the distribution chain. Who are these people? Well you have the lobsterman, his sternman, the lobster wholesaler, the lobster retailer, the lobster pound operators, and so forth. Without any doubt, the major mishandling of lobsters occurs aboard the lobster boat and at lobster pools where they are eventually sold as *"culls" to* wholesalers, retailers, and the retail customer.

Whether some people in the distribution chain are uncaring, unsympathetic, or untrained, the practice of mishandling lobsters and throwing them around like sticks and stones are unkind acts to say the very least. I have tried to be as objective as possible in the writing of this book, but my presence during the mishandling of lobsters leaves me with no alternative other than to tell it like it is – to tell it like I saw it – up close and personal!

If a lobster gives up one or both of its claws in a conflict with an adversary on the bottom of the ocean, then that is justifiable because it is akin with the ways of nature and the survival of the fittest. Landed individuals can't do anything about such circumstances. But personal experience has revealed random acts on the part of some lobstermen that have contributed to the cull lobster population. A common occurrence would be when a lobsterman was having a bad day. The weather was bad, the catch was off, he was becoming a little tired, and he was becoming a little short on patience. When these conditions are present, mistakes are made, things get rushed, and lobsters sometimes lose their claws. The situation usually arises when a trap is hauled up from the bottom, and during the sorting out process, a lobster will not yield a death grip that its large claws have on the netting of the trap. The impatient lobster fisherman will try to release the lobster's grip, and without trying to spread the claw sections apart, he might give the lobster a yank, and the claws are severed from the body of the lobster. That lobster then becomes a cull lobster. There have also been occasions when sub-legal size lobsters were thrown back into the ocean with such great momentum that injury was a very possible outcome. And I have witnessed firsthand the mistreatment of lobsters being removed from their holding tanks at lobster pools. Some lobsters were literally thrown into bags or cartons when getting ready for shipment or for the sale to a retail customer. One such happening was extremely bothersome to the point of protest: a young employee at one of Boston's largest shippers of live lobsters took a 1½ pound *"sleepy"* lobster out of a holding tank and threw it fifteen feet and into a hard plastic storage container. This rough handling of the lobster is, in essence, quite unnecessary when gentleness can accomplish the same purpose and with considerably less effort. The American lobster is, after all, an animal. It possesses many of the same functions as those of humans. It has a brain, a heart, a liver, a stomach, and many other organs. And of all the life forms of crustaceans, it is of high intelligence.

Having finally been caught, we have a marine specie that is now deposed on the kitchen counter or in the kitchen sink. There it remains virtually motionless with its ripper and crusher claws banded shut, and unknowing that it is about to be boiled, broiled, baked, or fried to death! When a lobster is to be boiled it is being put to death by being plunged head first into a pot of hot boiling water. When a lobster is about to be baked or broiled it is turned over onto its backside and slit with a sharp knife from end to end before being placed in a hot oven. What a way, after coping through the years, for a lobster to end up!

It seems the American lobster shows up in the darndest places, like in the movie "Tom Horn." In that movie, there is a scene depicting a multitude of people sitting at a long table in the middle of the prairie (of all places). And, on each place setting there are two red, boiled lobsters. Tom Horn, played by Steve McQueen, tries to eat a lobster for the first time, and experiences a certain amount of difficulty trying to get at the meat beneath the shell. After finally managing to crack the shell of one of the lobster's large claws, he manages to squirt everyone within striking distance with the juice from the lobster's claw. Everyone giggled!

Then, we should not forget the Red Lobster television commercials, one of which depicts a man running out of the ocean with a lobster held high in one of this outstretched hands. Such a happening is so unrealistic that it is somewhat insulting to our intelligence. Yet the lobster is used as an advertising attention-grabber and, in this regard, seems to be effective. It even gets my attention!

Robert Delano Martin

In the writing of this book, this has become bothersome to me. My wish for the lobster is that it will receive just and humane treatment by the many people who handle it prior to its consumption. The lobster deserves more respect than it has received. Therefore, I say: ***Be Merciful Unto The Lobster!!***

ABOUT THE AUTHOR

Robert Delano Martin is a long-time resident of Beverly, Massachusetts, has a wonderful marriage of more than forty-seven years and is the father of four grown children. He is a graduate of Northeastern University, is a veteran of the USAF and the Korean War epoch.

The author's familiarity with his subject – the American lobster – has been derived from close study of the animal through data research, by going out fishing with various lobstermen, and as an outcome of discussions with lobster marine scientists and other parties working with lobsters at the state and federal levels.

It is through these interactions and much time spent in and around the coastal lobster fishery that the author is able to bring the reader up close and personal to his subject: the American Lobster – Homarus americanus. .

For quite a period of time in the past, the author wrote a bi-weekly one-half page spread for the *Beverly Evening Times* entitled, "THEN AND NOW."

Printed in the United States
806900002B